Life Insurance Mathematics

Springer
Berlin
Heidelberg
New York
Barcelona
Budapest
Hong Kong
London
Milan
Paris
Santa Clara
Singapore
Tokyo

Hans U. Gerber

Life Insurance Mathematics

with exercises contributed by Samuel H. Cox

Third Edition 1997

 Springer

Swiss Association of Actuaries Zürich

Professor Hans U. Gerber
Ecole des H. E. C.
Université de Lausanne
CH-1015 Lausanne
Switzerland
e-mail: hgerber@hec.unil.ch

Professor Samuel H. Cox, F.S.A.
Georgia State University
Dept. of Risk Management
and Insurance
Atlanta, GA 30303-3083
USA
e-mail: insshc@panther.gsu.edu

Translator of the first edition:
Professor Walther Neuhaus
University of Oslo
Norway

CIP data applied for

Die Deutsche Bibliothek - CIP-Einheitsaufnahme

Gerber, Hans U.:
Life insurance mathematics / Hans U. Gerber. With exercises
contributed by Samuel H. Cox. Swiss Association of Actuaries,
Zürich. [Transl. of the first ed.: Walther Neuhaus]. - 3. ed. -
Berlin ; Heidelberg ; New York ; Barcelona ; Budapest ; Hong
Kong ; London ; Milan ; Paris ; Santa Clara ; Singapore ;
Tokyo : Springer, 1997
   Dt. Ausg. u.d.T.: Lebensversicherungsmathematik
   ISBN 3-540-62242-X

Mathematics Subject Classification (1991): 62P05

ISBN 3-540-62242-X Springer-Verlag Berlin Heidelberg New York

ISBN 3-540-58858-2 2nd edition Springer-Verlag Berlin Heidelberg New York

Typesetting: Camera-ready copy from the authors
SPIN 10563181   41/3143-5 4 3 2 1 0 – Printed on acid-free paper

To Cecil Nesbitt

Foreword

Halley's Comet has been prominently displayed in many newspapers during the last few months. For the first time in 76 years it appeared this winter, clearly visible against the nocturnal sky. This is an appropriate occasion to point out the fact that Sir Edmund Halley also constructed the world's first life table in 1693, thus creating the scientific foundation of life insurance. Halley's life table and its successors were viewed as deterministic laws, i.e. the number of deaths in any given group and year was considered to be a well defined number that could be calculated by means of a life table. However, in reality this number is random. Thus any mathematical treatment of life insurance will have to rely more and more on probability theory.

By sponsoring this monograph the Swiss Association of Actuaries wishes to support the "modern" probabilistic view of life contingencies. We are fortunate that Professor Gerber, an internationally renowned expert, has assumed the task of writing the monograph. We thank the Springer-Verlag and hope that this monograph will be the first in a successful series of actuarial texts.

Zürich, March 1986

Hans Bühlmann
President
Swiss Association of Actuaries

Preface

Two major developments have influenced the environment of actuarial mathematics. One is the arrival of powerful and affordable computers; the once important problem of numerical calculation has become almost trivial in many instances. The other is the fact that today's generation is quite familiar with probability theory in an intuitive sense; the basic concepts of probability theory are taught at many high schools. These two factors should be taken into account in the teaching and learning of actuarial mathematics. A first consequence is, for example, that a recursive algorithm (for a solution) is as useful as a solution expressed in terms of commutation functions. In many cases the calculations are easy; thus the question "why" a calculation is done is much more important than the question "how" it is done. The second consequence is that the somewhat embarrassing deterministic model can be abandoned; nowadays nothing speaks against the use of the stochastic model, which better reflects the mechanisms of insurance. Thus the discussion does not have to be limited to expected values; it can be extended to the deviations from the expected values, thereby quantifying the risk in the proper sense.

The book has been written in this spirit. It is addressed to the young reader (where "young" should be understood in the sense of operational time) who likes applied mathematics and is looking for an introduction into the basic concepts of life insurance mathematics.

In the first chapter an overview of the theory of compound interest is given. In Chapters 2–6 various forms of insurance and their mechanisms are discussed in the basic model. Here the key element is the future lifetime of a life aged x, which is denoted by T and which is (of course!) a random variable. In Chapter 7 the model is extended to multiple decrements, where different causes for departure (for example death and disability) are introduced. In Chapter 8 insurance policies are considered where the benefits are contingent on more than one life (for example widows' and orphans' pensions). In all these chapters the discussion focuses on a single policy, which is possible in the stochastic model, as opposed to the deterministic model, where each policy is considered as a member of a large group of identical policies. In Chapter 9 the risk arising from a group of policies (a *portfolio*) is examined. It is shown how the distribution of the aggregate claims can be calculated recursively. Information about this distribution is indispensable when reinsurance is purchased. The topic of Chapter 10 is of great practical importance; for simplicity of presentation the expense loading is considered only in this chapter. Chapter 11 examines some statistical problems, for instance, how to estimate the distribution of T from observations. The book has been written without much compromise; however, the appendix should be a sign of the conciliatory nature of the author. For the very same reason the basic probability space $(\Omega, \mathcal{F}, P)$ shall be mentioned at least once: now!

The publication of this book was made possible by the support of the Fund for the Encouragement of Actuarial Mathematics of the Swiss Association of Actuaries; my sincere thanks go to the members of its committee, not in the least for the freedom granted to me. I would like to thank in particular Professor Bühlmann and Professor Leepin for their valuable comments and suggestions. Of course I am responsible for any remaining flaws.

For some years now a team of authors has been working on a comprehensive text, which was commissioned by the Society of Actuaries and will be published in 1987 in its definitive form. The cooperation with the coauthors Professors Bowers, Hickman, Jones and Nesbitt has been an enormously valuable experience for me.

Finally I would like to thank my assistant, Markus Lienhard, for the careful perusal of the galley proofs and Springer-Verlag for their excellent cooperation.

Lausanne, March 1986   *Hans U. Gerber*

Acknowledgement

I am indebted to my colleague, Dr. Walther Neuhaus (University of Oslo), who translated the text into English and carried out the project in a very competent and efficient way. We are also very grateful to Professor Hendrik Boom (University of Manitoba) for his expert advice.

Lausanne and Winnipeg, April 1990   *Hans U. Gerber*

Acknowledgement

The second edition contains a rich collection of exercises, which have been prepared by Professor Samuel H. Cox of Georgia State University of Atlanta, who is an experienced teacher of the subject. I would like to express my sincere thanks to my American colleague: due to his contribution, the book will not only find *readers* but it will find *users*!

Lausanne, August 1995   *Hans U. Gerber*

Acknowledgement

The second edition has been sold out rapidly. This led to the present third edition, in which several misprints have been corrected. I am thankful to Sam Cox, Cheng Shixue, Wolfgang Quapp, André Dubey and Jean Cochet for their valuable advice.

At this occasion I would like to thank Springer and the Swiss Association of Actuaries for authorising the Chinese, Slovenian and Russian editions of Life Insurance Mathematics. I am indebted to Cheng Shixue, Yan Ying, Darko Medved and Valery Mishkin. From my own experience I know that translating a scientific text is a challenging task.

Lausanne, January 1997   *Hans U. Gerber*

Contents

Appendix C. Exercises

Appendix D. Solutions

Chapter 1. The Mathematics of Compound Interest

1.1 Mathematical Bases of Life Contingencies

To life insurance mathematics primarily two areas of mathematics are fundamental: the theory of compound interest and probability theory. This chapter gives an introduction to the first topic. The probabilistic model will be introduced in the next chapter; however, it is assumed that the reader is familiar with the basic principles of probability theory.

1.2 Effective Interest Rates

An interest rate is always stated in conjunction with a *basic time unit*; for example, one might speak of an annual rate of 6%. In addition, the *conversion period* has to be stated; this is the time interval at the end of which interest is credited or "compounded". An interest rate is called *effective* if the conversion period and the basic time unit are identical; in that case interest is credited at the end of the basic time unit.

Let i be an effective annual interest rate; for simplicity we assume that i is the same for all years. We consider an account (or fund) where the initial capital F_0 is invested, and where at the end of year k an additional amount of r_k is invested, for $k = 1, \cdots, n$. What is the balance at the end of n years? Let F_k be the balance at the end of year k, including the payment of r_k. Interest credited on the previous year's balance is iF_{k-1}. Thus

$$F_k = F_{k-1} + iF_{k-1} + r_k, \quad k = 1, \cdots, n. \tag{1.2.1}$$

We may write this recursive formula as

$$F_k - (1+i)F_{k-1} = r_k; \tag{1.2.2}$$

if we multiply this equation by $(1+i)^{n-k}$ and sum over all values of k, all but two terms on the left hand side vanish, and we obtain

$$F_n = (1+i)^n F_0 + \sum_{k=1}^{n}(1+i)^{n-k} r_k. \tag{1.2.3}$$

The powers of $(1+i)$ are called *accumulation factors*. The accumulated value of an initial capital C after h years is $(1+i)^h C$. Equation (1.2.3) illustrates an obvious result: the capital at the end of the interval is the accumulated value of the initial capital plus the sum of the accumulated values of the intermediate deposits.

The *discount factor* is defined as

$$v = \frac{1}{1+i}. \qquad (1.2.4)$$

Equation (1.2.3) can now be written as

$$v^n F_n = F_0 + \sum_{k=1}^{n} v^k r_k. \qquad (1.2.5)$$

Hence the present value of a capital C, due at time h, is $v^h C$.

If we write equation (1.2.1) as

$$F_k - F_{k-1} = i F_{k-1} + r_k \qquad (1.2.6)$$

and sum over k we obtain

$$F_n - F_0 = \sum_{k=1}^{n} i F_{k-1} + \sum_{k=1}^{n} r_k. \qquad (1.2.7)$$

Thus the increment of the fund is the sum of the total interest credited and the total deposits made.

1.3 Nominal Interest Rates

When the conversion period does not coincide with the basic time unit, the interest rate is called *nominal*. An annual interest rate of 6% with a conversion period of 3 months means that interest of $6\%/4 = 1.5\%$ is credited at the end of each quarter. Thus an initial capital of 1 increases to $(1.015)^4 = 1.06136$ at the end of one year. Therefore, an annual nominal interest rate of 6%, convertible quarterly, is equivalent to an annual effective interest rate of 6.136%.

Now, let i be a given annual effective interest rate. We define $i^{(m)}$ as the nominal interest rate, convertible m times per year, which is equivalent to i. Equality of the accumulation factors for one year leads to the equation

$$\left(1 + \frac{i^{(m)}}{m}\right)^m = 1 + i, \qquad (1.3.1)$$

which implies that

$$i^{(m)} = m[(1+i)^{1/m} - 1]. \qquad (1.3.2)$$

The limiting case $m \to \infty$ corresponds to continuous compounding. Let

$$\delta = \lim_{m \to \infty} i^{(m)} \; ; \tag{1.3.3}$$

this is called the *force of interest* equivalent to i. Writing (1.3.2) as

$$i^{(m)} = \frac{(1+i)^{1/m} - (1+i)^0}{1/m} \, , \tag{1.3.4}$$

we see that δ is the derivative of the function $(1+i)^x$ at the point $x = 0$. Thus we find that

$$\delta = \ln(1+i) \tag{1.3.5}$$

or

$$e^{\delta} = 1 + i \, . \tag{1.3.6}$$

We can verify this result by letting $m \to \infty$ in (1.3.1) and using the definition (1.3.3).

Thus the accumulation factor for a period of h years is $(1+i)^h = e^{\delta h}$; the discount factor for the same period of time is $v^h = e^{-\delta h}$. Here the length of the period h may be any real number.

Intuitively it is obvious that $i^{(m)}$ is a decreasing function of m. We can give a formal proof of this by interpreting $i^{(m)}$ as the slope of a secant, see (1.3.4), and using the convexity of the function $(1+i)^x$. The following numerical illustration is for $i = 6\%$.

m	$i^{(m)}$
1	0.06000
2	0.05913
3	0.05884
4	0.05870
6	0.05855
12	0.05841
∞	0.05827

1.4 Continuous Payments

We consider a fund as in Section 1.2, but now we assume that payments are made continuously with an annual instantaneous rate of payment of $r(t)$. Thus the amount deposited to the fund during the infinitesimal time interval from t to $t + dt$ is $r(t)\, dt$. Let $F(t)$ denote the balance of the fund at time t. We assume that interest is credited continuously, according to a, possibly time-dependent, force of interest $\delta(t)$. Interest credited in the infinitesimal time interval from t to $t + dt$ is $F(t)\delta(t)\,dt$. The total increase in the capital during this interval is thus

$$dF(t) = F(t)\delta(t)\,dt + r(t)\,dt\,. \tag{1.4.1}$$

To solve the corresponding differential equation

$$F'(t) = F(t)\delta(t) + r(t)\,, \tag{1.4.2}$$

we write

$$\frac{d}{dt}[e^{-\int_0^t \delta(s)\,ds}F(t)] = e^{-\int_0^t \delta(s)\,ds}r(t)\,. \tag{1.4.3}$$

Integration with respect to t from 0 to h gives

$$e^{-\int_0^h \delta(s)\,ds}F(h) - F(0) = \int_0^h e^{-\int_0^t \delta(s)\,ds}r(t)\,dt\,. \tag{1.4.4}$$

Thus the value at time 0 of a payment to be made at time t (i.e. its *present value*) is obtained by multiplication with the factor

$$e^{-\int_0^t \delta(s)\,ds}\,. \tag{1.4.5}$$

From (1.4.4) we further obtain

$$F(h) = e^{\int_0^h \delta(s)\,ds}F(0) + \int_0^h e^{\int_t^h \delta(s)\,ds}r(t)\,dt\,. \tag{1.4.6}$$

Thus the value at time h of a payment made at time $t < h$ (its accumulated value) is obtained by multiplication with the factor

$$e^{\int_t^h \delta(s)\,ds}\,. \tag{1.4.7}$$

In the case of a constant force of interest, i.e. $\delta(t) = \delta$, the factors (1.4.5) and (1.4.7) are reduced to the discount factors and accumulation factors introduced in Section 1.2.

1.5 Interest in Advance

Until now it was assumed that interest was to be credited at the end of each conversion period (or *in arrears*). But sometimes it is useful to assume that interest is credited at the beginning of each conversion period. Interest credited in this way is also referred to as *discount*, and the corresponding rate is called *discount rate* or *rate of interest-in-advance*.

Let d be an annual effective discount rate. A person investing an amount of C will be credited interest equal to dC immediately, and the invested capital

C will be returned at the end of the period. Investing the interest dC at the same conditions, the investor will receive additional interest of $d(dC) = d^2C$, and the additional invested amount will be returned at the end of the year; reinvesting the interest yields additional interest of $d(d^2C) = d^3C$, and so on. Repeating this process ad infinitum, we find that the investor will receive the total sum of

$$C + dC + d^2C + d^3C + \cdots = \frac{1}{1-d}C \tag{1.5.1}$$

at the end of the year in return for investing the initial capital C. The equivalent effective interest rate i is given by the equation

$$\frac{1}{1-d} = 1 + i, \tag{1.5.2}$$

which leads to

$$d = \frac{i}{1+i}. \tag{1.5.3}$$

This result has an obvious interpretation: if a capital of 1 unit is invested, d (the interest payable at the beginning of the year) is the discounted value of the interest i to be paid at the end of the year. Furthermore, (1.5.2) implies that

$$i = \frac{d}{1-d}. \tag{1.5.4}$$

Thus the interest payable at the end of the year is the accumulated value of the interest payable at the beginning of the year.

Let $d^{(m)}$ be the equivalent nominal rate of interest-in-advance credited m times per year. The investor thus obtains interest of $\frac{d^{(m)}}{m}C$ at the beginning of a conversion period, and his capital C is returned at the end of it. Equality of the accumulation factors for this mth part of a year is expressed by

$$\frac{1}{1 - d^{(m)}/m} = 1 + \frac{i^{(m)}}{m} = (1+i)^{1/m}. \tag{1.5.5}$$

This leads to

$$d^{(m)} = m[1 - (1+i)^{-1/m}]. \tag{1.5.6}$$

In analogy with (1.5.3) one obtains

$$d^{(m)} = \frac{i^{(m)}}{1 + i^{(m)}/m}, \tag{1.5.7}$$

resulting in a very simple relation between $i^{(m)}$ and $d^{(m)}$:

$$\frac{1}{d^{(m)}} = \frac{1}{m} + \frac{1}{i^{(m)}}. \tag{1.5.8}$$

It follows that

$$\lim_{m \to \infty} d^{(m)} = \lim_{m \to \infty} i^{(m)} = \delta, \tag{1.5.9}$$

which was to be expected: when interest is compounded continuously, the difference between interest in advance and interest in arrears vanishes.

The following numerical illustration is for $i = 6\%$.

m	$d^{(m)}$
1	0.05660
2	0.05743
3	0.05771
4	0.05785
6	0.05799
12	0.05813
∞	0.05827

1.6 Perpetuities

In this section we introduce certain types of perpetual payment streams (*perpetuities*) and calculate their present values. The resulting formulae are very simple and will later be useful for calculating the present value of annuities with a finite term.

First we consider perpetuities consisting of annual payments of 1 unit. If the first payment occurs at time 0, the perpetuity is called a *perpetuity-due*, and its present value is denoted by $\ddot{a}_{\overline{\infty}|}$. Thus

$$\ddot{a}_{\overline{\infty}|} = 1 + v + v^2 + \cdots = \frac{1}{1-v} = \frac{1}{d}. \tag{1.6.1}$$

If the first payment is made at the end of year 1, we call the perpetuity an *immediate perpetuity*. Its present value is denoted by $a_{\overline{\infty}|}$, and is given by

$$a_{\overline{\infty}|} = v + v^2 + v^3 + \cdots = \frac{v}{1-v} = \frac{1}{i}. \tag{1.6.2}$$

Let us now consider perpetuities where payments of $1/m$ are made m times each year. If the payments are made in advance (first payment of $1/m$ at time 0), the present value is denoted by $\ddot{a}_{\overline{\infty}|}^{(m)}$ and is

$$\ddot{a}_{\overline{\infty}|}^{(m)} = \frac{1}{m} + \frac{1}{m}v^{1/m} + \frac{1}{m}v^{2/m} + \cdots = \frac{1}{m}\frac{1}{1-v^{1/m}} = \frac{1}{d^{(m)}}, \tag{1.6.3}$$

cf. (1.5.6). If the payments are made in arrears (first payment of $1/m$ at time $1/m$), the present value is denoted by $a_{\overline{\infty}|}^{(m)}$ and given by

$$a_{\overline{\infty}|}^{(m)} = \frac{1}{m}v^{1/m} + \frac{1}{m}v^{2/m} + \frac{1}{m}v^{3/m} + \cdots$$

$$= \frac{1}{m} \frac{v^{1/m}}{1 - v^{1/m}} = \frac{1}{m[(1+i)^{1/m} - 1]}$$

$$= \frac{1}{i^{(m)}}, \tag{1.6.4}$$

cf. (1.3.2).

The results in (1.6.3) and (1.6.4) lead to an interpretation of the identity (1.5.8): since a perpetuity-due and an immediate perpetuity differ only by a payment of $1/m$ at time 0, their present values differ by $1/m$.

Let us now consider a continuous perpetuity with constant rate of payment $r = 1$ and starting at time 0. Its present value is denoted by $\bar{a}_{\overline{\infty}|}$ and given by

$$\bar{a}_{\overline{\infty}|} = \int_0^\infty e^{-\delta t} dt = \frac{1}{\delta}. \tag{1.6.5}$$

The same result can be obtained by letting $m \to \infty$ in (1.6.3) or (1.6.4).

The systematic pattern in formulae (1.6.1)–(1.6.5) is evident.

A certain type of perpetuities with increasing payments is defined by two parameters, m (the number of payments per year) and q (the number of increases per year); we assume that q is a factor of m. If for instance, $m = 12$ and $q = 4$, payments are made monthly and increase quarterly. In general, the payments of such an increasing perpetuity-due are defined as follows:

Time				Payment
0	$1/m$	$\cdots$	$1/q - 1/m$	$1/(mq)$
$1/q$	$1/q + 1/m$	$\cdots$	$2/q - 1/m$	$2/(mq)$
$2/q$	$2/q + 1/m$	$\cdots$	$3/q - 1/m$	$3/(mq)$
$3/q$	$3/q + 1/m$	$\cdots$	$4/q - 1/m$	$4/(mq)$
and so on				

In particular, the last m/q payments of year k are k/m each. We denote the present value of such a perpetuity by $(I^{(q)}\ddot{a})_{\overline{\infty}|}^{(m)}$. We can calculate it by representing the sequence of increasing payments as a sum of perpetuities with constant payments of $1/(mq)$ payable m times per year, and beginning at times $0, 1/q, 2/q, \cdots$. Thus we obtain the surprisingly simple formula

$$
\begin{aligned}
(I^{(q)}\ddot{a})_{\overline{\infty}|}^{(m)} &= \frac{1}{q} \ddot{a}_{\overline{\infty}|}^{(m)} [1 + v^{1/q} + v^{2/q} + \cdots] \\
&= \ddot{a}_{\overline{\infty}|}^{(m)} \ddot{a}_{\overline{\infty}|}^{(q)} \\
&= \frac{1}{d^{(m)}} \frac{1}{d^{(q)}}.
\end{aligned} \tag{1.6.6}
$$

The corresponding immediate annuity differs only in that each payment is made one mth year later, thus giving

$$(I^{(q)}a)_{\overline{\infty}|}^{(m)} = v^{1/m}\,(I^{(q)}\ddot{a})_{\overline{\infty}|}^{(m)} = \frac{v^{1/m}}{d^{(m)}}\frac{1}{d^{(q)}} = \frac{1}{i^{(m)}}\frac{1}{d^{(q)}}\,. \qquad (1.6.7)$$

A superscript of 1 is always omitted. For instance, the present value of a perpetuity-due with annual payments of $1, 2, \cdots$ is

$$(I\ddot{a})_{\overline{\infty}|} = (I^{(1)}\ddot{a})_{\overline{\infty}|}^{(1)} = \frac{1}{d^2}\,. \qquad (1.6.8)$$

Equations (1.6.6) and (1.6.7) may also be used with $m \to \infty$ to calculate present values of continuous payment streams. One obtains for instance

$$(\bar{I}\bar{a})_{\overline{\infty}|} = \int_0^\infty te^{-\delta t}dt = \frac{1}{\delta^2} \qquad (1.6.9)$$

and

$$(I\bar{a})_{\overline{\infty}|} = \int_0^\infty [t+1]e^{-\delta t}dt = \frac{1}{\delta d}\,, \qquad (1.6.10)$$

without actually calculating the integrals.

We conclude this section by considering a perpetuity with arbitrary annual payments of $r_0, r_1, r_2, \cdots$ (at times $0, 1, 2, \cdots$). Its present value, denoted simply by $\ddot{a}$, is

$$\ddot{a} = r_0 + vr_1 + v^2r_2 + \cdots\,. \qquad (1.6.11)$$

Such a variable perpetuity may be represented as a sum of constant perpetuities in the following way:

Annual payment	Starts at time
r_0	0
$r_1 - r_0$	1
$r_2 - r_1$	2
$r_3 - r_2$	3
and so on	

The present value of this perpetuity may therefore be expressed as

$$\ddot{a} = \frac{1}{d}\left\{r_0 + v(r_1 - r_0) + v^2(r_2 - r_1) + \cdots\right\}\,, \qquad (1.6.12)$$

which is useful if the differences of r_k are simpler than the r_k themselves. If, in particular, r_k is a polynomial in k, the present value $\ddot{a}$ may be calculated by repeated differencing. For instance, using $r_k = k + 1$ one may verify (1.6.8).

We can use (1.6.11) to calculate the present value of exponentially growing payments. Letting

$$r_k = e^{\tau k} \text{ for } k = 0, 1, 2, \cdots\,, \qquad (1.6.13)$$

one obtains

$$\ddot{a} = \frac{1}{1 - e^{-(\delta - \tau)}} , \qquad (1.6.14)$$

provided that $\tau < \delta$.

1.7 Annuities

In practice, annuities are more frequently encountered than perpetuities. An annuity is defined as a sequence of payments of a limited duration, which we denote by n. In what follows we consider some standard types of annuities, or annuities-certain as they sometimes are called.

The present value of an annuity-due with n annual payments of 1 starting at time 0, is denoted by $\ddot{a}_{\overline{n}|}$. It is given by

$$\ddot{a}_{\overline{n}|} = 1 + v + v^2 + \cdots + v^{n-1} . \qquad (1.7.1)$$

Representing the annuity as the difference of two perpetuities (one starting at time 0, the other at time n), we find that

$$\ddot{a}_{\overline{n}|} = \ddot{a}_{\overline{\infty}|} - v^n \ddot{a}_{\overline{\infty}|} = \frac{1}{d} - v^n \frac{1}{d} = \frac{1 - v^n}{d} . \qquad (1.7.2)$$

This result can be verified by directly evaluating the geometric sum (1.7.1).

In a similar way one obtains from (1.6.2),(1.6.3) and (1.6.4) the formulas

$$a_{\overline{n}|} = \frac{1 - v^n}{i} , \qquad (1.7.3)$$

$$\ddot{a}_{\overline{n}|}^{(m)} = \frac{1 - v^n}{d^{(m)}} , \qquad (1.7.4)$$

$$a_{\overline{n}|}^{(m)} = \frac{1 - v^n}{i^{(m)}} . \qquad (1.7.5)$$

Note that only the denominator varies, depending on the payment mode (immediate/due) and frequency. Note that n must be an integer in (1.7.2) and (1.7.3), and a multiple of $1/m$ in (1.7.4) and (1.7.5).

The final or accumulated value of annuities is also of interest. This is defined as the accumulated value of the payment stream at time n, and the usual symbol used is s. The final value is obtained by multiplying the initial value with the accumulation factor $(1 + i)^n$:

$$\ddot{s}_{\overline{n}|} = \frac{(1 + i)^n - 1}{d} , \qquad (1.7.6)$$

$$s_{\overline{n}|} = \frac{(1 + i)^n - 1}{i} , \qquad (1.7.7)$$

$$\ddot{s}_{\overline{n}|}^{(m)} = \frac{(1 + i)^n - 1}{d^{(m)}} , \qquad (1.7.8)$$

$$s_{\overline{n}|}^{(m)} = \frac{(1 + i)^n - 1}{i^{(m)}} . \qquad (1.7.9)$$

Another simple relation between the initial value and the final value of a constant annuity may easily be verified:

$$\frac{1}{a_{\overline{n}|}} = \frac{1}{s_{\overline{n}|}} + i .$$

(1.7.10)

Let us now consider an increasing annuity-due with parameters q and m:

Time				Payment
0	$1/m$	$\cdots$	$1/q - 1/m$	$1/(mq)$
$1/q$	$1/q + 1/m$	$\cdots$	$2/q - 1/m$	$2/(mq)$
$2/q$	$2/q + 1/m$	$\cdots$	$3/q - 1/m$	$3/(mq)$
$\cdots$				$\cdots$
$n - 1/q$	$n - 1/q + 1/m$	$\cdots$	$n - 1/m$	n/m

Such an increasing annuity can be represented as an increasing perpetuity starting at time 0, minus an identical increasing perpetuity starting at time n, minus a constant annuity starting at time n. Thus we may write

$$(I^{(q)}\ddot{a})_{\overline{n}|}^{(m)} = (I^{(q)}\ddot{a})_{\overline{\infty}|}^{(m)} - v^n \, (I^{(q)}\ddot{a})_{\overline{\infty}|}^{(m)} - v^n n \, \ddot{a}_{\overline{\infty}|}^{(m)} .$$

(1.7.11)

Substituting (1.6.6) and (1.6.3) and using (1.7.4), we obtain the equation

$$(I^{(q)}\ddot{a})_{\overline{n}|}^{(m)} = \frac{\ddot{a}_{\overline{n}|}^{(q)} - nv^n}{d^{(m)}} .$$

(1.7.12)

Similarly the present value of the corresponding immediate annuity is calculated:

$$(I^{(q)}a)_{\overline{n}|}^{(m)} = \frac{\ddot{a}_{\overline{n}|}^{(q)} - nv^n}{i^{(m)}} .$$

(1.7.13)

Note that in these equations n must be a multiple of $1/q$.

Important special cases are the combinations of $m = 1$ and $q = 1$, $m = 12$ and $q = 1$, $m = 12$ and $q = 12$, $m = \infty$ and $q = 1$, and $m = \infty$ and $q = \infty$. Equations (1.7.12) and (1.7.13) facilitate the evaluation of the present and final values for these combinations.

The annuities just considered are known as *standard increasing annuities* (I). *Standard decreasing annuities* (D) are similar, but the payments are made in the reversed order. The sum of a standard increasing annuity and its corresponding standard decreasing annuity is of course a constant annuity. This relation carries over to the present values, and we obtain

$$(I^{(q)}\ddot{a})_{\overline{n}|}^{(m)} + (D^{(q)}\ddot{a})_{\overline{n}|}^{(m)} = \left(n + \frac{1}{q} \right) \ddot{a}_{\overline{n}|}^{(m)} .$$

(1.7.14)

Using (1.7.12) and (1.7.14) and the identity

$$\ddot{a}_{\overline{n}|}^{(q)} = a_{\overline{n}|}^{(q)} + (1 - v^n)\frac{1}{q},$$
(1.7.15)

we obtain

$$(D^{(q)}\ddot{a})_{\overline{n}|}^{(m)} = \frac{n - a_{\overline{n}|}^{(q)}}{d^{(m)}}.$$
(1.7.16)

The direct derivation of this identity is also instructive: the standard decreasing annuity-due may be interpreted as a constant perpetuity with $m$thly payments of n/m, minus a series of deferred perpetuities-due, each with constant $m$thly payments of $1/(mq)$, and starting at times $1/q, 2/q, \cdots, n$.

1.8 Repayment of a Debt

Let S be the value at time 0 of a debt that is to be repaid by payments $r_1, \cdots, r_n$, made at the end of years $k = 1, 2, \cdots, n$. Then S must be the present value of the payments:

$$S = vr_1 + v^2 r_2 + \cdots + v^n r_n.$$
(1.8.1)

Let S_k be the principal outstanding, i.e. the remaining debt immediately after r_k has been paid. It consists of the previous year's debt, accumulated for one year, minus r_k:

$$S_k = (1 + i)S_{k-1} - r_k, \quad k = 1, \cdots, n.$$
(1.8.2)

This equation may be written as

$$r_k = iS_{k-1} + (S_{k-1} - S_k).$$
(1.8.3)

From (1.8.3) it is evident that each payment consists of two components, *interest* on the running debt and *reduction of principal*.

Substituting $-S_k$ for F_k, one sees that (1.8.2) is equivalent to (1.2.2). Thus all results of Section 1.2 carry over with the appropriate substitution. From (1.2.3) one obtains

$$S_k = (1 + i)^k S - \sum_{h=1}^{k} (1 + i)^{k-h} r_h,$$
(1.8.4)

and one may verify that $S_n = 0$, using (1.8.1). Similarly, (1.2.5) may be used to show

$$S_k = vr_{k+1} + v^2 r_{k+2} + \cdots + v^{n-k} r_n.$$
(1.8.5)

Formula (1.8.4) is the *retrospective formula*, and (1.8.5) is the *prospective formula* for the outstanding principal.

The payments $r_1, \cdots, r_k$ may be chosen arbitrarily, subject to the constraint (1.8.1). Some of the formulae in Section 1.7 may be derived by proper choice of the payment stream.

For instance, a debt of $S = 1$ can be repaid by the payments

$$r_1 = r_2 = \cdots = r_{n-1} = i, \ r_n = 1 + i. \tag{1.8.6}$$

In this case only interest is paid for first $n - 1$ years, and the entire debt, together with the last year's interest, is repaid at the end of the nth year. From (1.8.1) one finds

$$1 = i \, a_{\overline{n}|} + v^n, \tag{1.8.7}$$

which is another form of (1.7.3).

The debt of $S = 1$ may also be repaid by constant payments of

$$r_1 = r_2 = \cdots = r_n = \frac{1}{a_{\overline{n}|}}. \tag{1.8.8}$$

As an alternative to repaying the creditor at times $1, \cdots, n-1$, one could pay only the interest on S as in (1.8.6). In order to cover the final repayment one could make equal deposits to a fund that is to accumulate to 1 at the end of n years; from this it obvious that the annual deposit must be $1/\, s_{\overline{n}|}$. Since the total annual outgo must be the same in both cases, we arrive once again at equation (1.7.10).

Suppose now that we repay a debt of $S = n$ so that the principal outstanding decreases linearly to 0, $S_k = n - k$ for $k = 0, \cdots, n$. From (1.8.3) it is evident that $r_k = i(n - k + 1) + 1$. Using (1.8.1) one obtains the identity

$$n = i \, (Da)_{\overline{n}|} + a_{\overline{n}|}, \tag{1.8.9}$$

giving

$$(Da)_{\overline{n}|} = \frac{n - a_{\overline{n}|}}{i}. \tag{1.8.10}$$

This result is a special case ($m = q = 1$) of (1.7.16).

The loan itself may consist of a series of payments. Assume that equal payments of 1 are received by the debtor at times $0, 1, \cdots, n - 1$. At the end of each year interest on the received amounts is paid, and, in addition, the total amount received is repaid at time n:

$$r_k = ik \ \text{ for } k = 1, \cdots, n - 1 \ , \ r_n = in + n. \tag{1.8.11}$$

From the equality of the present values one obtains

$$\ddot{a}_{\overline{n}|} = i \, (Ia)_{\overline{n}|} + nv^n. \tag{1.8.12}$$

Equation (1.7.13) is obtained for the special case of $q = m = 1$.

Many other ways of repayment may be thought up. Present values of annuities-due can be derived if one assumes that interest is paid in advance. Another variant is the assumption that interest is debited m times a year, and that the debt is repaid q time a year in equal instalments (q a factor of m).

1.9 Internal Rate of Return

An investor pays a price p, which entitles him to n future payments. The payments are denoted by $r_1, \cdots, r_n$ and payment r_k is due at time τ_k, for $k = 1, \cdots, n$. What is the resulting rate of return?

The present value of the payment stream to be received by the investor is a function of the force of interest δ. Define

$$a(\delta) = \sum_{k=1}^{n} \exp(-\delta\tau_k)r_k \,. \tag{1.9.1}$$

Let t be the solution of the equation

$$a(t) = p \,. \tag{1.9.2}$$

The *internal rate of return* or *investment yield* is defined as $i = e^t - 1$.

Equation (1.9.2) may be solved by standard numerical methods, such as interval bisection or the Newton-Raphson method. We shall present a method which is more efficient than the former and simpler than the latter of those methods.

Consider the function

$$f(\delta) = \ln\left(a(\delta)/r\right), \tag{1.9.3}$$

(here $r = r_1 + \cdots + r_n$ denotes the undiscounted sum of the payments). It is not difficult to verify that

$$f(0) = 0, \qquad f'(\delta) = a'(\delta)/a(\delta) < 0 \,,$$
$$f''(\delta) = a''(\delta)/a(\delta) - (a'(\delta)/a(\delta))^2 > 0 \,. \tag{1.9.4}$$

(The last inequality may be verified by interpreting $f''(\delta)$ as a variance). Interpreting $f(\delta)/\delta$ as the slope of a secant and noting that f is a convex function by (1.9.4), we see that $f(\delta)/\delta$ is an increasing function of δ. Hence, for $0 < s < t < u$ one has the inequality

$$f(s)/s < f(t)/t < f(u)/u \,, \tag{1.9.5}$$

giving

$$\frac{f(t)}{f(s)}s < t < \frac{f(t)}{f(u)}u \,. \tag{1.9.6}$$

Thus we have proved that

$$\frac{\ln\left(p/r\right)}{\ln\left(a(s)/r\right)}s < t < \frac{\ln\left(p/r\right)}{\ln\left(a(u)/r\right)}u \,. \tag{1.9.7}$$

If one has a lower bound s and an upper bound u for the solution t of (1.9.2), these bounds may immediately be improved by (1.9.7).

The process may be iterated, yielding the following algorithm: Start with an arbitrary value δ_0, and calculate the values $\delta_1, \delta_2, \cdots$ by the recursive formula

$$\delta_{k+1} = \frac{\ln(p/r)}{\ln(a(\delta_k)/r)}\delta_k, \quad k = 0, 1, \cdots. \tag{1.9.8}$$

For $\delta_0 < t$ the resulting sequence will be monotone increasing and converge to t. For $\delta_0 > t$ the sequence will decrease monotonically to t. Thus any arbitrary positive value may be chosen for δ_0.

The method will be illustrated for a security (face amount Fr. 5000, yearly coupons Fr. 300). Assume that the security has been bought for Fr. 5250, for a remaining running time of 9 years. Thus we have

$$\begin{aligned}
p &= 5250, \\
\tau_k &= k \ (k = 1, \cdots, 9), \\
r_k &= 300 \ (k = 1, \cdots, 8), \\
r_9 &= 5300, \\
r &= 7700.
\end{aligned}$$

Assuming we know that the current yield of similar securities lies between 5% and 5.5%, we can use (1.9.7) with $s = \ln 1.05$ and $u = \ln 1.055$ to obtain improved bounds for the investment yield. We thus establish the bounds $0.051461 < t < 0.051572$, and obtain for $i = e^t - 1$:

$$5.2808\% < i < 5.2925\% \tag{1.9.9}$$

The algorithm defined by formula (1.9.8) may be used to obtain greater precision. In order to demonstrate its efficiency, we have chosen an artificially small initial value ($\delta_0 = \ln 1.01$ i.e. $i_0 = 1\%$) and an artificially large initial value ($\delta_0 = \ln 1.1$ i.e. $i_0 = 10\%$). The results have been compiled in the following table. In both cases the solution is arrived at in 4 iterations.

k	δ_k	i_k	δ_k	i_k
0	0.009950	0.01	0.095310	0.1
1	0.050612	0.051914	0.052627	0.054037
2	0.051503	0.052853	0.051551	0.052902
3	0.051524	0.052875	0.051525	0.052876
4	0.051525	0.052875	0.051525	0.052875

A sufficient condition for the existence of an internal rate of return as defined by (1.9.2), is that all payments r_k are positive. If some of the payments are negative (in practice this means that the investor has to supply additional capital), equation (1.9.2) may have several roots. The internal rate of return is not uniquely defined in such cases.

Chapter 2. The Future Lifetime
of a Life Aged x

2.1 The Model

Let us consider a person aged x years, also called a *life aged x* and denoted by (x). We denote his or her future lifetime by T or, more explicitly, by $T(x)$. Thus $x + T$ will be the age at death of the person.

The future lifetime T is a random variable with a probability distribution function

$$G(t) = \Pr\left(T \le t\right), \quad t \ge 0\,. \tag{2.1.1}$$

The function $G(t)$ represents the probability that the person will die within t years, for any fixed t. We assume that G, the probability distribution of T, is known. We also assume that G is continuous and has a probability density $g(t) = G'(t)$. Thus one may write

$$g(t)dt = \Pr\left(t < T < t + dt\right), \tag{2.1.2}$$

this being the probability that death will occur in the infinitesimal time interval from t to $t + dt$ (or that (x)'s age at death will fall between $x + t$ and $x + t + dt$).

Probabilities and expected values of interest may be expressed in terms of the functions g and G. Nevertheless, the international actuarial community uses a time-honoured notation, to which we shall adhere. For example, the probability that a life aged x will die within t years, is denoted by the symbol $_tq_x$. We have thus the relation

$$_tq_x = G(t)\,. \tag{2.1.3}$$

Similarly,

$$_tp_x = 1 - G(t) \tag{2.1.4}$$

denotes the probability that a life aged x will survive at least t years. Another commonly used symbol is

$$
\begin{aligned}
_{s|t}q_x &= \Pr\left(s < T < s + t\right) \\
&= G(s + t) - G(s) \\
&= {}_{s+t}q_x - {}_sq_x\,,
\end{aligned}
\tag{2.1.5}
$$

denoting the probability that the life aged x will survive s years and subsequently die within t years.

We denote by $_tp_{x+s}$ the conditional probability that the person will survive another t years, after having attained the age $x + s$. Thus

$$_tp_{x+s} = \Pr\left(T > s + t | T > s\right) = \frac{1 - G(s+t)}{1 - G(s)} . \tag{2.1.6}$$

Similarly, we define

$$_tq_{x+s} = \Pr\left(T \leq s + t | T > s\right) = \frac{G(s+t) - G(s)}{1 - G(s)} , \tag{2.1.7}$$

the conditional probability of dying within t years, given that the age of $x + s$ has been attained.

Identities in frequent use are

$$_{s+t}p_x = 1 - G(s+t) = [1 - G(s)]\frac{1 - G(s+t)}{1 - G(s)} = {_sp_x}\,{_tp_{x+s}} , \tag{2.1.8}$$

and

$$_{s|t}q_x = G(s+t) - G(s) = [1 - G(s)]\frac{G(s+t) - G(s)}{1 - G(s)} = {_sp_x}\,{_tq_{x+s}} . \tag{2.1.9}$$

These identities have an obvious interpretation.

The expected remaining lifetime of a life aged x is $\mathrm{E}(T)$, and denoted by $\overset{\circ}{e}_x$. Its definition is

$$\overset{\circ}{e}_x = \int_0^\infty tg(t)dt , \tag{2.1.10}$$

or, in terms of the distribution function,

$$\overset{\circ}{e}_x = \int_0^\infty [1 - G(t)]dt = \int_0^\infty {_tp_x}\,dt . \tag{2.1.11}$$

If $t = 1$, the index t is usually omitted in the symbols $_tq_x$, $_tp_x$, $_{s|t}q_x$. Thus q_x is the probability of dying within 1 year, and $_{s|}q_x$ is the probability of surviving s years and subsequently dying within 1 year.

2.2 The Force of Mortality

The *force of mortality* of (x) at the age $x + t$ is defined by

$$\mu_{x+t} = \frac{g(t)}{1 - G(t)} = -\frac{d}{dt}\ln\left[1 - G(t)\right] . \tag{2.2.1}$$

From (2.1.2) and (2.1.4) one may derive an alternative expression for the probability of dying in the interval between t and $t + dt$:

$$\Pr(t < T < t + dt) = {}_t p_x \mu_{x+t} dt.\tag{2.2.2}$$

The expected future lifetime of (x) can now be written as

$$\overset{\circ}{e}_x = \int_0^\infty t\, {}_t p_x \mu_{x+t} dt.\tag{2.2.3}$$

The approximation

$$_s q_{x+t} \approx \mu_{x+t} s\tag{2.2.4}$$

is valid for small values of s, as one may verify by exchanging the roles of s and t in (2.1.9) and comparing the result with (2.2.2).

The force of mortality may also be defined by

$$\mu_{x+t} = -\frac{d}{dt} \ln {}_t p_x.\tag{2.2.5}$$

Integration of (2.2.5) yields

$$_t p_x = e^{-\int_0^t \mu_{x+s} ds}.\tag{2.2.6}$$

2.3 Analytical Distributions of T

We call the function G an analytical or "mathematical" probability distribution if it may be expressed by a simple formula. There are different reasons for postulating an analytical distribution for T.

In the past efforts have been made to derive universally valid analytic expressions for $G(t)$ from certain basic postulates, in analogy with the laws of physics. These efforts, seen from a 20th century point of view, now seem rather naive and surrounded with a certain mystique.

An analytical formula has the advantage that $G(t)$ can readily be calculated from a small number of numeric parameters. Statistical inference in particular is facilitated when only a few parameters need to be estimated. This may be an important consideration when the available data are scarce.

Analytical formulae also have some attractive theoretical properties. Their popularity is akin to the popularity of the normal distribution in statistics: A normal model is often used, partly motivated by the Central Limit Theorem, but mainly for its mathematical tractability.

Some examples of analytical distributions follow, each bearing the name of its "inventor".

De Moivre (1724) postulated the existence of a maximum age ω for human beings and assumed that T was uniformly distributed between the ages of 0 and $\omega - x$, leading to $g(t) = \frac{1}{\omega - x}$ for $0 < t < \omega - x$. The force of mortality then becomes

$$\mu_{x+t} = \frac{1}{\omega - x - t}, \quad 0 < t < \omega - x, \tag{2.3.1}$$

which is an increasing function of t.

Gompertz (1824) postulated that the force of mortality would grow exponentially,

$$\mu_{x+t} = Bc^{x+t}, \quad t > 0 \tag{2.3.2}$$

which reflects the aging process better than De Moivre's law and in addition removes the assumption of a maximum age ω.

The law (2.3.2) was generalized by *Makeham* (1860), who postulated the law

$$\mu_{x+t} = A + Bc^{x+t}, \quad t > 0. \tag{2.3.3}$$

Makeham's mortality law adds a constant, age independent component $A > 0$ to the exponentially growing force of mortality of (2.3.2).

A special case of the mortality laws of Gompertz (by putting $c = 1$) and Makeham (by making $B = 0$) is that of a constant force of mortality. The probability distribution of T then becomes the exponential distribution. While mathematically very simple, this distribution does not reflect human mortality in a realistic way.

From (2.3.3) and (2.2.6), and putting $m = B / \ln c$, the survival probability in Makeham's model may be derived:

$$_t p_x = \exp\left(-At - mc^x(c^t - 1)\right). \tag{2.3.4}$$

Weibull (1939) suggested that the force of mortality grows as a power of t, instead of exponentially:

$$\mu_{x+t} = k(x + t)^n, \tag{2.3.5}$$

with the fixed parameters $k > 0$ and $n > 0$. The survival probability then becomes

$$_t p_x = \exp\left(-\frac{k}{n+1}\left[(x+t)^{n+1} - x^{n+1}\right]\right). \tag{2.3.6}$$

2.4 The Curtate Future Lifetime of (x)

We now return to the general model introduced in Sections 2.1 and 2.2 and define the random variables $K = K(x)$, $S = S(x)$, $S^{(m)} = S^{(m)}(x)$, all closely related to the original random variable T.

We define $K = [T]$, the number of completed future years lived by (x), or the *curtate future lifetime* of (x). The probability distribution of the integer-valued random variable K is given by

$$\Pr(K = k) = Pr(k \leq T < k + 1) = {_k p_x}\, q_{x+k} \tag{2.4.1}$$

for $k = 0, 1, \cdots$. The expected value of K is called the expected curtate future lifetime of (x) and is denoted by e_x. Thus

$$e_x = \sum_{k=1}^{\infty} k \Pr(K = k) = \sum_{k=1}^{\infty} k \, {}_k p_x \, q_{x+k} \tag{2.4.2}$$

or

$$e_x = \sum_{k=1}^{\infty} \Pr(K \geq k) = \sum_{k=1}^{\infty} {}_k p_x \,. \tag{2.4.3}$$

Use of the expected curtate lifetime has the advantage that (2.4.1) and (2.4.2) are easier to evaluate than (2.1.11) and (2.2.3). Another advantage is that one only needs the distribution of K in order to find e_x.

Let S be the fraction of a year during which (x) is alive in the year of death, i.e.

$$T = K + S \,. \tag{2.4.4}$$

The random variable S has a continuous distribution between 0 and 1. Approximating its expected value by $\frac{1}{2}$ we find, from (2.4.4), the approximation

$$\overset{\circ}{e}_x \approx e_x + \frac{1}{2} \,, \tag{2.4.5}$$

which may be used in practice for the expected future lifetime of (x).

Let us assume that K and S are independent random variables, so that the conditional distribution of S, given K, is independent of K; thus

$$\Pr(S \leq u | K = k) = \frac{{}_u q_{x+k}}{q_{x+k}} \tag{2.4.6}$$

will not depend on the argument k, so that one can write

$${}_u q_{x+k} = H(u) \, q_{x+k} \tag{2.4.7}$$

for $k = 0, 1, \cdots$ and $0 \leq u \leq 1$, and some function $H(u)$.

If we assume that $H(u) = u$ (uniform distribution between 0 and 1), then the approximation (2.4.5) is exact. Moreover, using (2.4.4) and the assumed independence, the variance of T becomes

$$\text{Var}(T) = \text{Var}(K) + \frac{1}{12} \,. \tag{2.4.8}$$

For positive integers m we define the random variable

$$S^{(m)} = \frac{1}{m} [mS + 1] \,. \tag{2.4.9}$$

Thus $S^{(m)}$ is derived from S by rounding to the next higher multiple of $1/m$. The distribution of $S^{(m)}$ has its mass in the points $\frac{1}{m}, \frac{2}{m}, \cdots, 1$. Note that independence between K and S implies independence between K and $S^{(m)}$. Furthermore, if S has a uniform distribution between 0 and 1, then $S^{(m)}$ has a discrete uniform distribution.

2.5 Life Tables

In the previous sections of this chapter we considered a person of age x. The probability distribution of his future lifetime can be constructed by adopting a suitable *life table*.

A life table is essentially a table of one-year death probabilities q_x, which completely defines the distribution of K. In the next section we will show how to approximate the distribution of T by interpolation in the life table.

Life tables are constructed from statistical data (see Chapter 11). The construction of a life table involves estimation, graduation and extrapolation techniques (the latter are used to account for changing mortality patterns over time).

Life tables are constructed for certain population groups, differentiated by factors such as sex, race, generation and insurance type. The initial age x can have a significant influence in such tables. For instance, let x denote the age when the person bought life insurance. Since insurance is only offered to individuals of good health (sometimes only after a medical test), it is reasonable to expect that a person who has just bought insurance, will be of better health than a person who bought insurance several years ago, other factors (particularly age) being equal. This phenomenon is taken into account by *select life tables*. In a select life table, the probabilities of death are graded according to the age at entry. Thus $q_{[x]+t}$ is the one-year probability of death for $(x+t)$ with x as entry age. Selection leads to the inequalities

$$q_{[x]} < q_{[x-1]+1} < q_{[x-2]+2} < \cdots . \tag{2.5.1}$$

The selection effect has usually worn off after some years, say r years after entry. We assume that

$$q_{[x-r]+r} = q_{[x-r-1]+r+1} = q_{[x-r-2]+r+2} = \cdots = q_x . \tag{2.5.2}$$

The period r is called the *select period*, and the table used after the select period has expired, is called an *ultimate life table*.

Consider a person who buys a life insurance policy at age x. With a select period of 3 years, the following probabilities are needed in order to determine the distribution of K:

$$q_{[x]}, \ q_{[x]+1}, \ q_{[x]+2}, \ q_{x+3}, \ q_{x+4}, \ q_{x+5}, \cdots . \tag{2.5.3}$$

If a life table varies only with the attained age x, it is called an *aggregate life table*. It has the advantage of being single-entry, while a select life table is double-entry. The one-year probability of death at a given attained age in an aggregate life table will typically be a weighted average of the corresponding probabilities in the select life table and in the ultimate life table.

Though it is easy to use a select life table, cf. (2.5.3), we shall, for simplicity, use the notation of the aggregate life table in the sequel.

2.6 Probabilities of Death for Fractions of a Year

The distribution of K and its related quantities may be calculated from a life table. For example,

$$_k p_x = p_x\, p_{x+1}\, p_{x+2} \cdots p_{x+k-1}, \quad k = 1, 2, 3, \cdots, \tag{2.6.1}$$

cf. (2.1.8). To obtain the distribution of T by interpolation, assumptions are made regarding the pattern of the probabilities of death, $_u q_x$, or the force of mortality, μ_{x+u}, at intermediate ages $x + u$ (x an integer and $0 < u < 1$).
    We shall discuss three such assumptions.

Assumption a: Linearity of $_u q_x$

If one assumes that $_u q_x$ is a linear function of u, interpolation between $u = 0$ and $u = 1$ yields

$$_u q_x = u\, q_x . \tag{2.6.2}$$

We have seen in Section 2.4 that this is the case where K and S are independent, and S is uniformly distributed between 0 and 1. Then

$$_u p_x = 1 - u\, q_x \tag{2.6.3}$$

and (2.2.5) gives

$$\mu_{x+u} = \frac{q_x}{1 - u\, q_x} . \tag{2.6.4}$$

Assumption b: μ_{x+u} constant

A popular assumption is that the force of mortality is constant over each unit interval. Let us denote the constant value of μ_{x+u}, $(0 < u < 1)$ by $\mu_{x+\frac{1}{2}}$. Using (2.2.5) one finds

$$\mu_{x+\frac{1}{2}} = -\ln p_x . \tag{2.6.5}$$

It also follows that

$$_u p_x = e^{-u\mu_{x+\frac{1}{2}}} = (p_x)^u . \tag{2.6.6}$$

From (2.4.6) one derives

$$\Pr(S \le u | K = k) = \frac{1 - p_{x+k}^u}{1 - p_{x+k}} . \tag{2.6.7}$$

The conditional distribution of S, given $K = k$, is thus a truncated exponential distribution, and it depends on k. The random variables S and K are not independent in this case.

Assumption c: Linearity of $_{1-u}q_{x+u}$

This hypothesis, well-known in North America as the *Balducci assumption*, states

$$_{1-u}q_{x+u} = (1 - u)\, q_x \,.$$ (2.6.8)

This leads to

$$_u p_x = \frac{p_x}{_{1-u}p_{x+u}} = \frac{1 - q_x}{1 - (1 - u)\, q_x} \,.$$ (2.6.9)

From this and (2.2.5) we obtain

$$\mu_{x+u} = \frac{q_x}{1 - (1 - u)\, q_x} \,,$$ (2.6.10)

and finally

$$\Pr\left(S \le u | K = k\right) = \frac{u}{1 - (1 - u)\, q_{x+k}} \,.$$ (2.6.11)

This shows that the random variables S and K are not independent under the Balducci hypothesis,

Under each of the three assumptions the force of mortality is discontinuous at integer values. More embarrassing is the fact that under the Balducci assumption the force of mortality decreases between consecutive integers, cf. (2.6.10).

For $q_{x+k} \to 0$ both (2.6.7) and (2.6.11) converge to u. Thus, if the probabilities of death are small, S is "approximately" uniformly distributed and independent of K (even under assumptions b or c).

Chapter 3. Life Insurance

3.1 Introduction

Under a life insurance contract the benefit insured consists of a single payment, the *sum insured*. The time and amount of this payment may be functions of the random variable T that has been introduced in Chapter 2. Thus the time and amount of the payment may be random variables themselves.

The present value of the payment is denoted by Z; it is calculated on the basis of a fixed rate of interest i (the *technical* rate of interest). The expected present value of the payment, $E(Z)$, is the *net single premium* of the contract. This premium, however, does not in any way reflect the risk to be carried by the insurer. In order to assess this one requires further characteristics of the distribution of the random variable Z, for example its variance.

3.2 Elementary Insurance Types

3.2.1 Whole Life and Term Insurance

Let us consider a *whole life insurance*; this provides for payment of 1 unit at the end of the year of death. In this case the amount of the payment is fixed, while the time of payment $(K + 1)$ is random. Its present value is

$$Z = v^{K+1}. \tag{3.2.1}$$

The random variable Z ranges over the values $v, v^2, v^3, \cdots$, and the distribution of Z is determined by (3.2.1) and the distribution of K:

$$\Pr\left(Z = v^{k+1}\right) = \Pr\left(K = k\right) = {}_k p_x \, q_{x+k} \tag{3.2.2}$$

for $k = 0, 1, 2, \cdots$. The net single premium is denoted by A_x and given by

$$A_x = E[v^{K+1}] = \sum_{k=0}^{\infty} v^{k+1} \, {}_k p_x \, q_{x+k}. \tag{3.2.3}$$

The variance of Z may be calculated by the identity

$$\operatorname{Var}(Z) = E(Z^2) - A_x^2. \tag{3.2.4}$$

Replacing v by $e^{-\delta}$ we see that

$$E(Z^2) = E[e^{-2\delta(K+1)}] , \tag{3.2.5}$$

which is the net single premium calculated at twice the original force of interest. Thus calculating the variance is no more difficult than calculating the net single premium.

An insurance which provides for payment only if death occurs within n years is known as a *term insurance* of duration n. For example 1 unit is payable only if death occurs during the first n years, the actual time of payment still being the end of the year of death. One has

$$Z = \begin{cases} v^{K+1} & \text{for } K = 0, 1, \cdots, n-1 , \\ 0 & \text{for } K = n, n+1, n+2, \cdots . \end{cases} \tag{3.2.6}$$

The net single premium is denoted by $A^1_{x:\overline{n}|}$. It is

$$A^1_{x:\overline{n}|} = \sum_{k=0}^{n-1} v^{k+1} \,_kp_x \, q_{x+k} . \tag{3.2.7}$$

Again the second moment $E(Z^2)$ equals the net single premium at twice the original force of interest, as is seen from

$$Z^2 = \begin{cases} e^{-2\delta(K+1)} & \text{for } K = 0, 1, \cdots, n-1 , \\ 0 & \text{for } K = n, n+1, n+2, \cdots . \end{cases} \tag{3.2.8}$$

3.2.2 Pure Endowments

A *pure endowment* of duration n provides for payment of the sum insured only if the insured is alive at the end of n years:

$$Z = \begin{cases} 0 & \text{for } K = 0, 1, \cdots, n-1 , \\ v^n & \text{for } K = n, n+1, n+2, \cdots . \end{cases} \tag{3.2.9}$$

The net single premium is denoted by $A_{x:\overline{n}|}^{1}$ and is given by

$$A_{x:\overline{n}|}^{1} = v^n \,_np_x . \tag{3.2.10}$$

The formula for the variance of a Bernoulli random variable gives

$$\text{Var}(Z) = v^{2n} \,_np_x \,_nq_x . \tag{3.2.11}$$

3.2.3 Endowments

Assume that the sum insured is payable at the end of the year of death, if this occurs within the first n years, otherwise at the end of the nth year:

$$Z = \begin{cases} v^{K+1} & \text{for } K = 0, 1, \cdots, n-1 \ , \\ v^n & \text{for } K = n, n+1, n+2, \cdots \ . \end{cases} \qquad (3.2.12)$$

The net single premium is denoted by $A_{x:\overline{n}|}$. Denoting the present value of (3.2.6) by Z_1, and that of (3.2.9) by Z_2, one may obviously write

$$Z = Z_1 + Z_2 \ . \qquad (3.2.13)$$

As a consequence,

$$A_{x:\overline{n}|} = A^1_{x:\overline{n}|} + A_{x:\overline{n}|}^{1} \qquad (3.2.14)$$

and

$$\text{Var}(Z) = \text{Var}(Z_1) + 2\,\text{Cov}(Z_1, Z_2) + \text{Var}(Z_2) \ . \qquad (3.2.15)$$

The product $Z_1 Z_2$ is always zero, hence

$$\text{Cov}(Z_1, Z_2) = \text{E}(Z_1 Z_2) - \text{E}(Z_1)\text{E}(Z_2) = -\,A^1_{x:\overline{n}|} A_{x:\overline{n}|}^{1} \ . \qquad (3.2.16)$$

The variance of Z is thus given by

$$\text{Var}(Z) = \text{Var}(Z_1) + \text{Var}(Z_2) - 2\,A^1_{x:\overline{n}|} A_{x:\overline{n}|}^{1} \ . \qquad (3.2.17)$$

As a consequence of the last identity, the risk in selling an endowment policy, measured by the variance, is less than that in selling a term insurance to one person and a pure endowment to another.

So far, for simplicity, we have assumed a sum insured of 1. If the sum insured is C, then the net single premium is obtained by multiplying with C, and the variance by multiplying with C^2.

Let us finally consider an m year *deferred whole life insurance*. Its present value is

$$Z = \begin{cases} 0 & \text{for } K = 0, 1, \cdots, m-1 \ , \\ v^{K+1} & \text{for } K = m, m+1, m+2, \cdots \ . \end{cases} \qquad (3.2.18)$$

The net single premium is denoted by $_{m|}A_x$. Alternative formulae for its net single premium are

$$_{m|}A_x = {}_mp_x v^m\, A_{x+m} \qquad (3.2.19)$$

and

$$_{m|}A_x = A_x - A^1_{x:\overline{m}|} \ . \qquad (3.2.20)$$

The second moment $\text{E}(Z^2)$ again equals the net single premium at twice the original force of interest.

3.3 Insurances Payable at the Moment of Death

In the previous section it was assumed that the sum insured was payable at the end of the year of death. This assumption does not reflect insurance practice in a realistic way, but has the advantage that the formulae may be evaluated directly from a life table.

Let us now assume that the sum insured becomes payable at the instant of death, i.e. at time T. The present value of a payment of 1 payable immediately on death is

$$Z = v^T . \tag{3.3.1}$$

The net single premium is denoted by $\bar{A}_x$. Using (2.2.2) we find that

$$\bar{A}_x = \int_0^\infty v^t \, {}_tp_x\mu_{x+t}dt . \tag{3.3.2}$$

A practical approximation may be derived under *Assumption a* of Section 2.6. Writing

$$T = K + S = (K + 1) - (1 - S), \tag{3.3.3}$$

and making use of the assumed independence of K and S, as well as the uniform distribution of S, so that

$$E[(1 + i)^{1-S}] = \int_0^1 (1 + i)^u du = \bar{s}_{\overline{1}|} = \frac{i}{\delta} , \tag{3.3.4}$$

we find

$$\bar{A}_x = E[v^{K+1}] \, E[(1 + i)^{1-S}] = \frac{i}{\delta} A_x . \tag{3.3.5}$$

Thus the calculation of $\bar{A}_x$ is a simple extension of that of A_x.

A similar formula may be derived for term insurances. For endowments the factor i/δ is only used in the term insurance part:

$$
\begin{aligned}
\bar{A}_{x:\overline{n}|} &= \bar{A}^1_{x:\overline{n}|} + A_{x:\overline{n}|}^{\;\;1} \\
&= \frac{i}{\delta} A^1_{x:\overline{n}|} + A_{x:\overline{n}|}^{\;\;1} \\
&= A_{x:\overline{n}|} + \left(\frac{i}{\delta} - 1\right) A^1_{x:\overline{n}|} .
\end{aligned}
\tag{3.3.6}
$$

Let us finally assume that the sum insured is payable at the end of the mth part of the year in which death occurs, i.e. time $K + S^{(m)}$ in the notation of Section 2.4. The present value of a whole life insurance of 1 unit then becomes

$$Z = v^{K+S^{(m)}} . \tag{3.3.7}$$

For calculation of the net single premium we again use the *Assumption a* of Section 2.6. We write

$$K + S^{(m)} = (K + 1) - (1 - S^{(m)}) \tag{3.3.8}$$

in (3.3.7) and use the assumed independence of K and $S^{(m)}$, as well as the equation

$$E[(1+i)^{1-S^{(m)}}] = s_{\overline{1}|}^{(m)} = \frac{i}{i^{(m)}} \,. \tag{3.3.9}$$

Then we obtain

$$A_x^{(m)} = E[v^{K+1}]E[(1+i)^{1-S^{(m)}}] = \frac{i}{i^{(m)}} \, A_x \,. \tag{3.3.10}$$

Equation (3.3.5) may be verified by letting $m \to \infty$ in (3.3.10).

3.4 General Types of Life Insurance

We commence by considering a life insurance with benefits varying from year to year, and we assume that the sum insured is payable at the end of the year of death. If c_j denotes the sum insured during the jth year after policy issue, we have

$$Z = c_{K+1} v^{K+1} \,. \tag{3.4.1}$$

The distribution of Z and, in particular, the net single premium and higher moments are easy to calculate:

$$E[Z^h] = \sum_{k=0}^{\infty} c_{k+1}^h v^{h(k+1)} \, {}_k p_x \, q_{x+k} \,. \tag{3.4.2}$$

The insurance described may be represented as a combination of deferred life insurances, each of which has a constant sum insured. Thus the net single premium may be calculated in the following way:

$$E(Z) = c_1 \, A_x + (c_2 - c_1) \, {}_{1|}A_x + (c_3 - c_2) \, {}_{2|}A_x + \cdots \,. \tag{3.4.3}$$

In the case that the insurance covers only a term of n years, i.e. when $c_{n+1} = c_{n+2} = \cdots = 0$, the insurance may also be represented as a combination of term insurances starting immediately:

$$E(Z) = c_n \, A_{x:\overline{n}|}^1 + (c_{n-1} - c_n) \, A_{x:\overline{n-1}|}^1 + (c_{n-2} - c_{n-1}) \, A_{x:\overline{n-2}|}^1 + \cdots \,. \tag{3.4.4}$$

The alternative representations (3.4.3) and (3.4.4) are useful in calculating the net single premium, but not the higher order moments of Z.

If an insurance is payable immediately on death, the sum insured may in general be a function $c(t)$, $t \geq 0$, and we have

$$Z = c(T)v^T \,. \tag{3.4.5}$$

The net single premium is

$$E(Z) = \int_0^{\infty} c(t)v^t \, {}_t p_x \mu_{x+t} dt \,. \tag{3.4.6}$$

The actual calculation of the net single premium may be reduced to a calculation in the discrete model, see (3.4.2) with $h = 1$. From

$$
\begin{aligned}
\mathrm{E}(Z) &= \sum_{k=0}^{\infty} \mathrm{E}[Z|K = k]\Pr(K = k) \\
&= \sum_{k=0}^{\infty} \mathrm{E}[c(k + S)v^{k+S}|K = k]\Pr(K = k) \\
&= \sum_{k=0}^{\infty} \mathrm{E}[c(k + S)(1 + i)^{1-S}|K = k]v^{k+1}\Pr(K = k), \quad (3.4.7)
\end{aligned}
$$

we obtain

$$
\mathrm{E}(Z) = \sum_{k=0}^{\infty} c_{k+1}v^{k+1} \, {}_kp_x \, q_{x+k}, \tag{3.4.8}
$$

by defining

$$
c_{k+1} = \mathrm{E}[c(k + S)(1 + i)^{1-S}|K = k]. \tag{3.4.9}
$$

The conditional distribution of S, given $K = k$, is needed in order to evaluate the expression (3.4.9). Two assumptions about mortality at fractional ages are appropriate for making this evaluation.

Assumption a of Section 2.6 gives

$$
c_{k+1} = \int_0^1 c(k + u)(1 + i)^{1-u}du, \tag{3.4.10}
$$

whereas *Assumption b* of the same section results in

$$
c_{k+1} = \int_0^1 c(k + u)(1 + i)^{1-u} \frac{\mu_{x+k+\frac{1}{2}} \, p_{x+k}^u}{1 - p_{x+k}} \, du. \tag{3.4.11}
$$

As an illustration, consider the case of an exponentially increasing sum insured, $c(t) = e^{\tau t}$. This reduces formula (3.4.10) to

$$
c_{k+1} = e^{\tau k} \frac{e^{\delta} - e^{\tau}}{\delta - \tau}. \tag{3.4.12}
$$

Note that $\tau = 0$ gives us (3.3.5) back. The alternative formula (3.4.11) results in

$$
c_{k+1} = e^{\tau k} \frac{\mu_{x+k+\frac{1}{2}}}{1 - p_{x+k}} \frac{e^{\delta} - p_{x+k}e^{\tau}}{\delta + \mu_{x+k+\frac{1}{2}} - \tau}. \tag{3.4.13}
$$

(If the denominator in (3.4.12) or (3.4.13) should vanish, the quotients become e^{δ}. This will happen if the integrand in (3.4.10) or (3.4.11), respectively, is independent of u).

3.5 Standard Types of Variable Life Insurance

We begin by considering standard types where the sum insured is payable at the end of the year of death. The net single premium may be readily calculated and is useful also when the sum insured is payable immediately on death.

Let us consider a *standard increasing* whole life insurance, with $c_j = j$. The present value of the insurance is

$$Z = (K+1)v^{K+1}. \tag{3.5.1}$$

The net single premium is denoted by $(IA)_x$ and is given by

$$(IA)_x = \sum_{k=0}^{\infty} (k+1)v^{k+1} \, {}_kp_x \, q_{x+k}. \tag{3.5.2}$$

For the corresponding n-year term insurance we have

$$Z = \begin{cases} (K+1)v^{K+1} & \text{for } K = 0, 1, \cdots, n-1 \\ 0 & \text{for } K = n, n+1, n+2, \cdots \end{cases} \tag{3.5.3}$$

Its net single premium is denoted by $(IA)^1_{x:\overline{n}|}$ and may be obtained by limiting the summation in (3.5.2) to the first n terms. Inspired by (3.4.3) and (3.4.4) we may write

$$(IA)^1_{x:\overline{n}|} = A_x + {}_{1|}A_x + \cdots + {}_{n-1|}A_x - n \, {}_{n|}A_x \tag{3.5.4}$$

and

$$(IA)^1_{x:\overline{n}|} = n \, A^1_{x:\overline{n}|} - A^1_{x:\overline{n-1}|} - A^1_{x:\overline{n-2}|} - \cdots - A^1_{x:\overline{1}|}. \tag{3.5.5}$$

Note the difference between the symbols $(IA)^1_{x:\overline{n}|}$ and $(IA)_{x:\overline{n}|}$ - the latter being equal to the sum of the former and the net single premium for a pure endowment of n.

The benefits of a *standard decreasing* term insurance decrease linearly from n to 0, hence

$$Z = \begin{cases} (n-K)v^{K+1} & \text{for } K = 0, 1, \cdots, n-1 \\ 0 & \text{for } K = n, n+1, n+2, \cdots \end{cases} \tag{3.5.6}$$

Standard decreasing insurance is commonly used to guarantee repayment of a loan, provided that the debt outstanding also decreases linearly under the amortisation plan of the loan. The identities

$$(DA)^1_{x:\overline{n}|} = \sum_{k=0}^{n-1} (n-k)v^{k+1} \, {}_kp_x \, q_{x+k} \tag{3.5.7}$$

and

$$(DA)^1_{x:\overline{n}|} = A^1_{x:\overline{n}|} + A^1_{x:\overline{n-1}|} + A^1_{x:\overline{n-2}|} + \cdots + A^1_{x:\overline{1}|} \tag{3.5.8}$$

are obvious.

Let us now assume that the sum insured is payable immediately on death, i.e. Z is of the form (3.4.5), with some function $c(t)$. For these insurances we shall use *Assumption a* of Section 2.6 throughout this section.

If the sum insured is incremented annually, we have $c(t) = [t+1]$, hence

$$Z = (K+1)v^T . \tag{3.5.9}$$

The net single premium is denoted by $(I\bar{A})_x$. Calculating the expectation of

$$Z = (K+1)v^{K+1}(1+i)^{1-S} \tag{3.5.10}$$

and using the assumed independence of K and S as well as (3.3.4), we obtain the practical formula

$$(I\bar{A})_x = \frac{i}{\delta}(IA)_x . \tag{3.5.11}$$

Let us now consider the situation where the sum payable is incremented q times a year, by $1/q$ each time:

$$Z = (K + S^{(q)})v^T . \tag{3.5.12}$$

The corresponding net single premium is denoted by $(I^{(q)}\bar{A})_x$. Note that (3.5.12) may be rewritten as

$$Z = (K+1)v^T - v^T + S^{(q)}(1+i)^{1-S}v^{K+1} . \tag{3.5.13}$$

In computing the net single premium we use independence and the relation

$$\mathrm{E}[S^{(q)}(1+i)^{1-S}] = (I^{(q)}\bar{s})_{\overline{1}|} = \frac{\ddot{s}^{(q)}_{\overline{1}|} - 1}{\delta} = \frac{i - d^{(q)}}{d^{(q)}\delta} . \tag{3.5.14}$$

Hence we obtain

$$(I^{(q)}\bar{A})_x = (I\bar{A})_x - \bar{A}_x + \frac{i - d^{(q)}}{d^{(q)}\delta}A_x . \tag{3.5.15}$$

Substituting from (3.3.5) and (3.5.11), we find

$$(I^{(q)}\bar{A})_x = \frac{i}{\delta}(IA)_x - \frac{i}{\delta}A_x + \frac{i - d^{(q)}}{d^{(q)}\delta}A_x ; \tag{3.5.16}$$

This last expression may be evaluated directly.

In the case of a continuously increasing sum insured, $c(t) = t$, the present value is

$$Z = Tv^T , \tag{3.5.17}$$

and the net single premium

$$(\bar{I}\bar{A})_x = \frac{i}{\delta}(IA)_x - \frac{i}{\delta}A_x + \frac{i-\delta}{\delta^2}A_x \qquad (3.5.18)$$

is obtained by taking the limit $q \to \infty$ in (3.5.16).

The formulae (3.5.11), (3.5.16) and (3.5.18) may also be obtained by substituting the appropriate function $c(t)$ in (3.4.10). As an example, taking $c(t) = t$ leads to

$$c_{k+1} = \int_0^1 (k+u)(1+i)^{1-u}du = k\,\bar{s}_{\overline{1}|} + (\bar{I}\bar{s})_{\overline{1}|} = k\frac{i}{\delta} + \frac{i-\delta}{\delta^2}, \qquad (3.5.19)$$

which gives us (3.5.18).

Similar equations hold for the corresponding term insurances, for example

$$(I^{(q)}\bar{A})^1_{x:\overline{n}|} = \frac{i}{\delta}(IA)^1_{x:\overline{n}|} - \frac{i}{\delta}A^1_{x:\overline{n}|} + \frac{i-d^{(q)}}{d^{(q)}\delta}A^1_{x:\overline{n}|}. \qquad (3.5.20)$$

Obtaining an elegant derivation of (3.5.20) from (3.5.16) is left to the reader.

Finally we consider an n-year continuous term insurance with an initial sum insured of n, which is reduced q times a year, by $1/q$ each time:

$$Z = \begin{cases} (n+1/q - K - S^{(q)})v^T & \text{for } T < n \\ 0 & \text{for } T \geq n \end{cases}. \qquad (3.5.21)$$

This insurance may obviously be represented as the difference between a term insurance with constant sum of $n+1/q$ insured, and a term insurance with increasing sum insured. The net single premium is given by

$$(D^{(q)}\bar{A})^1_{x:\overline{n}|} = \left(n+\frac{1}{q}\right)\bar{A}^1_{x:\overline{n}|} - (I^{(q)}\bar{A})^1_{x:\overline{n}|}. \qquad (3.5.22)$$

3.6 Recursive Formulae

Recursion formulae may be used to write algorithms, but they also have interesting theoretical implications.

We start by considering a whole life insurance of 1 payable at the end of the year of death. One obviously has the equation

$$A_x = v\,q_x + v\,A_{x+1}\,p_x. \qquad (3.6.1)$$

Thus the values of A_x can be found recursively, starting with the highest possible age. The recursive equation may be proved algebraically by substitution of

$$_kp_x = p_x\,_{k-1}p_{x+1} \qquad (3.6.2)$$

in all but the first term of the summation (3.2.3). A probabilistic proof may
be built on the relation

$$E[v^{K+1}] = v \Pr(K = 0) + v E[v^K | K \geq 1] \Pr(K \geq 1). \qquad (3.6.3)$$

The interpretation of (3.6.1) is instructive. The net single premium at age x
is the expected value of a random variable defined as discounted sum insured
in case of death, and discounted net single premium at age $x + 1$ in case of
survival.

Another interpretation is evident if we write (3.6.1) as

$$A_x = v \, A_{x+1} + v(1 - A_{x+1}) \, q_x. \qquad (3.6.4)$$

First the amount of A_{x+1} is reserved in any case (death or survival). In case
of death an additional $1 - A_{x+1}$ is needed to cover the payment. The net
single premium of a one-year term insurance of this amount is $v(1 - A_{x+1}) \, q_x$.

Applying (3.6.4) at age $x + k$ we obtain

$$A_{x+k} - v \, A_{x+k+1} = v(1 - A_{x+k+1}) \, q_{x+k}, \quad k = 0, 1, 2 \cdots. \qquad (3.6.5)$$

Multiplying the above equation by v^k and summing over all values of k we
obtain

$$A_x = \sum_{k=0}^{\infty} v^k v(1 - A_{x+k+1}) \, q_{x+k}, \qquad (3.6.6)$$

so that the net single premium at age x is evidently the sum of the net single
premiums of a series of one-year term insurances.

Equation (3.6.4) may also be rewritten as

$$d \, A_{x+1} = (A_{x+1} - A_x) + v(1 - A_{x+1}) \, q_x. \qquad (3.6.7)$$

Thus the interest earned has a dual effect: On the one hand it increases the
net single premium (from age x to age $x + 1$), and on the other it finances a
fictitious one-year term insurance.

The continuous counterpart to a recursion formula is a differential equation. Consider the function $\bar{A}_x$, the expected value of v^T. For $h > 0$ we
have

$$\begin{aligned}
\bar{A}_x &= E[v^T | T \leq h] \Pr(T \leq h) + E[v^T | T > h] \Pr(T > h) \\
&= E[v^T | T \leq h] \, {}_h q_x + v^h \, \bar{A}_{x+h} \, {}_h p_x.
\end{aligned} \qquad (3.6.8)$$

Hence

$$\bar{A}_{x+h} - \bar{A}_x = (1 - v^h \, {}_h p_x) \, \bar{A}_{x+h} - E[v^T | T \leq h] \, {}_h q_x. \qquad (3.6.9)$$

Division by h and letting $h \to 0$ yields

$$\frac{d}{dx} \, \bar{A}_x = (\delta + \mu_x) \, \bar{A}_x - \mu_x. \qquad (3.6.10)$$

This equation can be recast in a form similar to (3.6.7):

$$\delta \, \bar{A}_x = \frac{d}{dx} \, \bar{A}_x + \mu_x (1 - \bar{A}_x) \, . \tag{3.6.11}$$

The differential equation has a similar interpretation as (3.6.7) for an infinitesimal time interval, which is seen by multiplying (3.6.11) by dt.

Only the two simplest types of insurance have been formally discussed in this section. The interpretations we have given for the recursion formulae resp. differential equations above are, of course, also valid for the general case and may therefore be used to derive the corresponding recursion formulae and differential equations.

Chapter 4. Life Annuities

4.1 Introduction

A life annuity consists of a series of payments which are made while the beneficiary (of initial age x) lives. Thus a life annuity may be represented as an annuity-certain with a term dependent on the remaining lifetime T. Its present value thus becomes a random variable, which we shall denote by Y.

The net single premium of a life annuity is its expected present value, $E(Y)$. More generally, the distribution of Y may also be of interest, as well as its moments.

A life annuity may, on the one hand, be the benefit of an insurance policy as a combination of pure endowments; on the other hand, periodic payment of premiums can also be considered as a life annuity, of course with the algebraic sign reversed.

4.2 Elementary Life Annuities

We consider a *whole life annuity-due* which provides for annual payments of 1 unit as long as the beneficiary lives. Payments are made at the time points $0, 1, \cdots, K$. The present value of this payment stream is

$$Y = 1 + v + v^2 + \cdots + v^K = \ddot{a}_{\overline{K+1}|} ; \qquad (4.2.1)$$

the probability distribution of this random variable is given by

$$\Pr\left(Y = \ddot{a}_{\overline{k+1}|}\right) = \Pr\left(K = k\right) = {}_k p_x \, q_{x+k} , \quad k = 0, 1, 2, \cdots . \qquad (4.2.2)$$

The net single premium, denoted by $\ddot{a}_x$, is the expected value of (4.2.1):

$$\ddot{a}_x = \sum_{k=0}^{\infty} \ddot{a}_{\overline{k+1}|} \, {}_k p_x \, q_{x+k} . \qquad (4.2.3)$$

The present value (4.2.1) may also be expressed as

$$Y = \sum_{k=0}^{\infty} v^k I_{\{K \geq k\}} , \qquad (4.2.4)$$

where I_A is the indicator function of an event A. The expectation of (4.2.4) is

$$\ddot{a}_x = \sum_{k=0}^{\infty} v^k {}_k p_x \, . \tag{4.2.5}$$

Thus we have found two expressions for the net single premium of a whole life annuity-due. In expression (4.2.3) we consider the whole annuity as a unit, while in (4.2.5) we think of the annuity as a series of pure endowments.

The net single premium may also be expressed in terms of the net single premium for a whole life insurance, the latter being given by (3.2.1) and (3.2.3). By virtue of (1.7.2), the net single premium (4.2.1) equals

$$Y = \frac{1 - v^{K+1}}{d} = \frac{1 - Z}{d} \, . \tag{4.2.6}$$

(This formula may also be obtained by viewing the life annuity as the difference of two perpetuities-due, one starting at time 0, the other at time $K + 1$.) Taking expectations yields

$$\ddot{a}_x = \frac{1 - A_x}{d} \, . \tag{4.2.7}$$

After transforming this identity to

$$1 = d \, \ddot{a}_x + A_x \, , \tag{4.2.8}$$

we may interpret it in terms of a debt of 1 unit with annual interest in advance, and a final payment of 1 unit at the end of the year of death. Of course the higher order moments of Y may also be derived from (4.2.6), so that, for instance,

$$\mathrm{Var}(Y) = \frac{\mathrm{Var}(Z)}{d^2} \, . \tag{4.2.9}$$

The present value of an n-year temporary life annuity-due is

$$Y = \begin{cases} \ddot{a}_{\overline{K+1}|} & \text{for } K = 0, 1, \cdots, n - 1 \, , \\ \ddot{a}_{\overline{n}|} & \text{for } K = n, n + 1, n + 2, \cdots \, . \end{cases} \tag{4.2.10}$$

Similarly to (4.2.3) and (4.2.5) the net single premium can be expressed by either

$$\ddot{a}_{x:\overline{n}|} = \sum_{k=0}^{n-1} \ddot{a}_{\overline{k+1}|} \, {}_k p_x \, q_{x+k} + \ddot{a}_{\overline{n}|} \, {}_n p_x \tag{4.2.11}$$

or

$$\ddot{a}_{x:\overline{n}|} = \sum_{k=0}^{n-1} v^k \, {}_k p_x \, . \tag{4.2.12}$$

Now we have

$$Y = \frac{1 - Z}{d} \, , \tag{4.2.13}$$

but here Z is defined by (3.2.12). As a consequence,

$$\ddot{a}_{x:\overline{n}|} = \frac{1 - A_{x:\overline{n}|}}{d}, \tag{4.2.14}$$

or

$$1 = d\,\ddot{a}_{x:\overline{n}|} + A_{x:\overline{n}|}. \tag{4.2.15}$$

The corresponding immediate life annuities provide for payments at times $1, 2, \cdots, K$:

$$Y = v + v^2 + \cdots + v^K = a_{\overline{K}|}. \tag{4.2.16}$$

The random variables (4.2.1) and (4.2.16) differ only by the constant term 1. Thus the net single premium a_x is given by

$$a_x = \ddot{a}_x - 1. \tag{4.2.17}$$

From equation (1.8.7), with $n = K + 1$, we obtain

$$1 = i\,a_{\overline{K}|} + (1+i)v^{K+1}. \tag{4.2.18}$$

Taking expectations yields

$$1 = i\,a_x + (1+i)\,A_x, \tag{4.2.19}$$

in analogy to (4.2.8).

The present value of an m year deferred life annuity-due with annual payments of 1 unit is

$$Y = \begin{cases} 0 & \text{for } K = 0, 1, \cdots, m-1, \\ v^m + v^{m+1} + \cdots + v^K & \text{for } K = m, m+1, \cdots. \end{cases} \tag{4.2.20}$$

The net single premium may be obtained from either one of the obvious relations:

$$_{m|}\ddot{a}_x = {}_mp_x v^m\,\ddot{a}_{x+m}, \tag{4.2.21}$$

$$_{m|}\ddot{a}_x = \ddot{a}_x - \ddot{a}_{x:\overline{m}|}. \tag{4.2.22}$$

4.3 Payments made more Frequently than Once a Year

Consider the case where payments of $1/m$ are made m times a year, i.e. at times $0, 1/m, 2/m, \cdots$, as long as the beneficiary, initially aged x, is alive. The net single premium of such an annuity is denoted by $\ddot{a}_x^{(m)}$. In analogy with (4.2.8) we have

$$1 = d^{(m)}\,\ddot{a}_x^{(m)} + A_x^{(m)}. \tag{4.3.1}$$

Hence we obtain

$$\ddot{a}_x^{(m)} = \frac{1}{d^{(m)}} - \frac{1}{d^{(m)}} A_x^{(m)} . \tag{4.3.2}$$

The equation may be interpreted in the following way: The life annuity payable m times a year can be viewed as the difference of two perpetuities, one starting at time 0, the other at time $K + S^{(m)}$. Taking expectations then yields (4.3.2).

To obtain expressions for $\ddot{a}_x^{(m)}$ in terms of $\ddot{a}_x$ we use again *Assumption a* of Section 2.6, so that (3.3.10) allows us to express $A_x^{(m)}$ of (4.3.2) in terms of A_x; if we then replace A_x in turn by $1 - d\,\ddot{a}_x$, (4.3.2) becomes

$$\ddot{a}_x^{(m)} = \frac{di}{d^{(m)}i^{(m)}} \ddot{a}_x - \frac{i - i^{(m)}}{d^{(m)}i^{(m)}} . \tag{4.3.3}$$

Introducing

$$\alpha(m) = \frac{di}{d^{(m)}i^{(m)}} \quad \text{and} \quad \beta(m) = \frac{i - i^{(m)}}{d^{(m)}i^{(m)}} , \tag{4.3.4}$$

we can then write (4.3.2) more economically as

$$\ddot{a}_x^{(m)} = \alpha(m)\,\ddot{a}_x - \beta(m) . \tag{4.3.5}$$

For $i = 5\%$ the coefficients $\alpha(m)$ and $\beta(m)$ are tabulated below, with $m = 12$ (monthly payments) and with $m = \infty$ (continuous payments).

m	$\alpha(m)$	$\beta(m)$
12	1.000197	0.46651
∞	1.000198	0.50823

Practical approximations in frequent use are

$$\alpha(m) \approx 1 , \quad \beta(m) \approx \frac{m - 1}{2m} . \tag{4.3.6}$$

These approximations are obtained from the Taylor expansion of the coefficients around $\delta = 0$, viz.

$$\alpha(m) = 1 + \frac{m^2 - 1}{12m^2}\delta^2 + \cdots , \tag{4.3.7}$$

$$\beta(m) = \frac{m - 1}{2m} + \frac{m^2 - 1}{6m^2}\delta + \cdots . \tag{4.3.8}$$

Apparently these approximations are useful only when the force of interest is sufficiently small.

The net single premium of a temporary life annuity-due with $m$thly payments can now also be calculated with the help of $\alpha(m)$ and $\beta(m)$:

$$
\begin{aligned}
\ddot{a}^{(m)}_{x:\overline{n}|} &= \ddot{a}^{(m)}_x - {}_np_x v^n \ddot{a}^{(m)}_{x+n} \\
&= \alpha(m)\,\ddot{a}_x - \beta(m) - {}_np_x v^n \left\{ \alpha(m)\,\ddot{a}_{x+n} - \beta(m) \right\} \\
&= \alpha(m)\,\ddot{a}_{x:\overline{n}|} - \beta(m)\left\{ 1 - {}_np_x v^n \right\}. \tag{4.3.9}
\end{aligned}
$$

The net single premium of an immediate life annuity (payments in arrears) may be calculated in terms of the corresponding life annuity-due:

$$
a^{(m)}_{x:\overline{n}|} = \ddot{a}^{(m)}_{x:\overline{n}|} - \frac{1}{m}\{ 1 - {}_np_x v^n \}. \tag{4.3.10}
$$

Let us now return to the calculation of $\ddot{a}^{(m)}_x$. Equations (4.2.8) and (4.3.1) give the exact expression

$$
\ddot{a}^{(m)}_x = \frac{d}{d^{(m)}}\ddot{a}_x - \frac{1}{d^{(m)}}\{ A^{(m)}_x - A_x \}, \tag{4.3.11}
$$

which may be interpreted in the following way: The life annuity on the left hand side provides payments of $1/m$ at times $0, 1/m, \cdots, K + S^{(m)} - 1/m$; it may be represented as the difference of two temporary annuities, the first providing payments at times $0, 1/m, \cdots, K + 1 - 1/m$, the second providing payments at times $K + S^{(m)}, K + S^{(m)} + 1/m, \cdots, K + 1 - 1/m$. This second temporary annuity may in turn be viewed as the difference of two perpetuities (one starting at $K + S^{(m)}$, the other at $K + 1$). The first temporary annuity has the same present value as an annuity-due which provides $K + 1$ annual payments of $d/d^{(m)}$. Taking expectations of the present values then yields (4.3.11).

Under *Assumption a*, we may use equation (3.3.10), giving

$$
\ddot{a}^{(m)}_x = \frac{d}{d^{(m)}}\ddot{a}_x - \beta(m)\,A_x; \tag{4.3.12}
$$

this formula has an obvious interpretation, which is not the case with the mathematically equivalent formula (4.3.5).

4.4 Variable Life Annuities

We start by considering a life annuity which provides payments of $r_0, r_1, r_2, \cdots$ at the time points $0, 1, \cdots, K$. The present value is

$$
Y = \sum_{k=0}^{\infty} v^k r_k I_{\{K \geq k\}}, \tag{4.4.1}
$$

and the net single premium

$$E(Y) = \sum_{k=0}^{\infty} v^k r_k \, {}_kp_x \tag{4.4.2}$$

may be readily calculated.

Take now a general life annuity with payments of $z_0, z_{1/m}, z_{2/m}, \cdots$ at time points $0, 1/m, 2/m, \cdots, K+S^{(m)}-1/m$. We start by replacing the m payments of each year by one advance payment with the same present value:

$$r_k = \sum_{j=0}^{m-1} v^{j/m} z_{k+j/m} \, , \quad k = 0, 1, 2, \cdots . \tag{4.4.3}$$

The correction term in the year of death amounts to a negative life insurance, the sum insured at time $k + u$, $0 < u < 1$ being the present value of the omitted payments:

$$c(k+u) = \sum_{j \in J} v^{j/m-u} z_{k+j/m} \, ; \tag{4.4.4}$$

here $J = J(u)$ is the set of those $j \in \{1, 2, \cdots, m-1\}$ for which $j/m > u$. In order to calculate the net single premium we use *Assumption a* of Section 2.6 and proceed along the lines of Section 3.4. Substituting (4.4.4) in equation (3.4.10) we obtain

$$\begin{aligned}
c_{k+1} &= \int_0^1 \sum_J (1+i)^{1-j/m} z_{k+j/m} du \\
&= \frac{1}{m} \sum_{j=1}^{m-1} j (1+i)^{1-j/m} z_{k+j/m} \, .
\end{aligned} \tag{4.4.5}$$

The net single premium for a general life annuity with payments m times a year is thus

$$\sum_{k=0}^{\infty} v^k r_k \, {}_kp_x - \sum_{k=0}^{\infty} c_{k+1} v^{k+1} \, {}_kp_x \, q_{x+k} \, , \tag{4.4.6}$$

with the coefficients defined in (4.4.3) and (4.4.5).

The case of a continuously payable annuity is obtained by letting $m \to \infty$. Let the payment rate at time t be $r(t)$. The present value is

$$Y = \int_0^T v^t r(t) \, dt \, . \tag{4.4.7}$$

The net single premium

$$E(Y) = \int_0^{\infty} v^t r(t) \, {}_tp_x dt \tag{4.4.8}$$

may be evaluated by (4.4.6), with coefficients

$$r_k = \int_0^1 v^u r(k+u)\, du, \tag{4.4.9}$$

$$c_{k+1} = \int_0^1 u(1+i)^{1-u} r(k+u)\, du. \tag{4.4.10}$$

We illustrate the point by a continuous life annuity with exponential growth ,

$$r(t) = e^{\tau t}. \tag{4.4.11}$$

From (4.4.9) and (4.4.10) we obtain

$$r_k = \frac{1 - e^{\tau - \delta}}{\delta - \tau} e^{\tau k} \tag{4.4.12}$$

and

$$c_{k+1} = \frac{e^{\delta - \tau} - 1 - (\delta - \tau)}{(\delta - \tau)^2} e^{\tau(k+1)} \tag{4.4.13}$$

for $\tau \neq \delta$, and

$$r_k = e^{\delta k}, \quad c_{k+1} = \frac{1}{2} e^{\delta(k+1)} \tag{4.4.14}$$

for $\tau = \delta$. In the case of a constant payment rate ($\tau = 0$), (4.4.12) and (4.4.13) become simply

$$r_k = \frac{d}{\delta}, \quad c_{k+1} = \beta(\infty), \tag{4.4.15}$$

which is in accordance with (4.3.12).

4.5 Standard Types of Life Annuity

Consider a life annuity of the form (4.4.1) with $r_k = k+1$. Its net single premium, which we denote by $(I\ddot{a})_x$, may be readily calculated by means of (4.4.2).

A simple identity connects $(I\ddot{a})_x$ and $(IA)_x$. Replacing n by $K+1$ in the identity

$$\ddot{a}_{\overline{n}|} = d\,(I\ddot{a})_{\overline{n}|} + nv^n, \tag{4.5.1}$$

see (1.8.12), and taking expectations we obtain

$$\ddot{a}_x = d\,(I\ddot{a})_x + (IA)_x, \tag{4.5.2}$$

which reminds us of (4.2.8).

We consider the case of m payments a year with annual increments:

$$z_{k+j/m} = \frac{k+1}{m}, \quad j = 0, 1, \cdots, m-1. \tag{4.5.3}$$

The net single premium of this life annuity is denoted by $(I\ddot{a})_x^{(m)}$. Representing this annuity as a sum of deferred annuities, we obtain, with (4.3.5)

$$
\begin{aligned}
(I\ddot{a})_x^{(m)} &= \sum_{k=0}^{\infty} {}_k p_x v^k \, \ddot{a}_{x+k}^{(m)} \\
&= \sum_{k=0}^{\infty} {}_k p_x v^k \{\alpha(m) \, \ddot{a}_{x+k} - \beta(m)\} \\
&= \alpha(m) \sum_{k=0}^{\infty} {}_k p_x v^k \, \ddot{a}_{x+k} - \beta(m) \sum_{k=0}^{\infty} {}_k p_x v^k \\
&= \alpha(m) \, (I\ddot{a})_x - \beta(m) \, \ddot{a}_x .
\end{aligned}
\tag{4.5.4}
$$

This expression may be evaluated directly.

Letting $m \to \infty$ we obtain the corresponding continuous annuity with payment rate $r(t) = [t+1]$. Its net single premium is given by

$$
\begin{aligned}
(I\bar{a})_x &= \int_0^{\infty} [t+1] v^t \, {}_t p_x dt \\
&= \alpha(\infty) \, (I\ddot{a})_x - \beta(\infty) \, \ddot{a}_x .
\end{aligned}
\tag{4.5.5}
$$

The present value of a continuous life annuity with payment rate $r(t) = t$ is

$$
Y = \int_0^T t v^t dt = (\bar{I}\bar{a})_{\overline{T}|} = \frac{\bar{a}_{\overline{T}|} - T v^T}{\delta} .
\tag{4.5.6}
$$

Taking expectations yields the formula

$$
(\bar{I}\bar{a})_x = \frac{\bar{a}_x - (\bar{I}\bar{A})_x}{\delta} .
\tag{4.5.7}
$$

This expression may be evaluated using (3.5.18) and (4.3.5) with $m = \infty$.

The derivation of the corresponding formulae for standard decreasing life and temporary annuities is left to the reader.

4.6 Recursion Formulae

We shall restrict our discussion to recursion formulae for the function $\ddot{a}_x$. Replacing ${}_k p_x$ by $p_x \, {}_{k-1} p_{x+1}$ in all except the first term in (4.2.5) we find

$$
\ddot{a}_x = 1 + v \, \ddot{a}_{x+1} \, p_x .
\tag{4.6.1}
$$

The values of $\ddot{a}_x$ may be calculated successively, starting with the highest possible age.

An equivalent expression is

$$
\ddot{a}_x = 1 + v \, \ddot{a}_{x+1} - v \, \ddot{a}_{x+1} \, q_x .
\tag{4.6.2}
$$

The net single premium is seen to cover the payment due at age x and the present value of the net single premium at age $x+1$, less the expected mortality gain.

Application of (4.6.2) at age $x + k$ yields

$$\ddot{a}_{x+k} - v\,\ddot{a}_{x+k+1} = 1 - v\,\ddot{a}_{x+k+1}\,q_{x+k}\,. \tag{4.6.3}$$

We multiply this equation by v^k and sum over k to obtain

$$\ddot{a}_x = \ddot{a}_{\overline{\infty}|} - \sum_{k=0}^{\infty} v^{k+1}\,\ddot{a}_{x+k+1}\,q_{x+k}\,. \tag{4.6.4}$$

The net single premium may thus be viewed as the present value of a perpetuity, reduced each year by the expected mortality gain.

Finally we can write (4.6.2) as

$$d\,\ddot{a}_{x+1} = 1 + (\,\ddot{a}_{x+1} - \ddot{a}_x\,) - v\,\ddot{a}_{x+1}\,q_x\,, \tag{4.6.5}$$

from which the role of the earned interest becomes evident.

In analogy with (4.6.5) one may derive the differential equation

$$\delta\,\bar{a}_x = 1 + \frac{d}{dx}\,\bar{a}_x - \mu_x\,\bar{a}_x \tag{4.6.6}$$

by substituting

$$\bar{A}_x = 1 - \delta\,\bar{a}_x\,, \quad \frac{d}{dx}\,\bar{A}_x = -\delta\frac{d}{dx}\,\bar{a}_x \tag{4.6.7}$$

in (3.6.11).

4.7 Inequalities

The net single premium $\bar{a}_x$ is occasionally confused with the present value $\bar{a}_{\overline{\mathring{e}_x}|}$. The values are different; in fact one has the inequality

$$\bar{a}_x < \bar{a}_{\overline{\mathring{e}_x}|}\,. \tag{4.7.1}$$

In view of (4.6.7) and the identity $v^t = 1 - \delta\,\bar{a}_{\overline{t}|}$, with $t = \mathring{e}_x$, an equivalent inequality may be found:

$$\bar{A}_x > v^{\mathring{e}_x}\,. \tag{4.7.2}$$

Each of these inequalities is a direct consequence of Jensen's inequality; for instance the second inequality means

$$E(v^T) > v^{E(T)},$$ (4.7.3)

which is obvious since v^t is a convex function of t.

In what follows we shall generalise these inequalities. Consider the net single premium $\bar{A}_x$ as a function of the force of interest δ:

$$\bar{A}_x(\delta) = E[e^{-\delta T}];$$ (4.7.4)

this is the Laplace transform of the distribution of T. We also define the function

$$f(\delta) = \{E[e^{-\delta T}]\}^{1/\delta}, \quad \delta > 0.$$ (4.7.5)

For small values of δ one may approximate (4.7.4) by $1 - \delta\, \mathring{e}_x$. Thus $\lim_{\delta \to 0} f(\delta)$ exists, and has the value

$$f(0) = \exp(-\mathring{e}_x).$$ (4.7.6)

Lemma: *The function $f(\delta)$ is monotone increasing.*

To prove the lemma we take two positive numbers $u < w$, and demonstrate that

$$f(w) > f(u).$$ (4.7.7)

Jensen's inequality implies

$$E[e^{-wT}] = E[\{e^{-uT}\}^{w/u}] > \{E[e^{-uT}]\}^{w/u}.$$ (4.7.8)

Hence

$$f(w)^w > f(u)^w,$$ (4.7.9)

from which (4.7.7) follows. This proves the lemma.

The lemma implies that $f(\delta) > f(0)$, hence

$$f(\delta)^\delta > f(0)^\delta.$$ (4.7.10)

From (4.7.6) one may derive the inequality (4.7.2) once more.

An interesting application uses three different forces of interest, $\delta_1 < \delta < \delta_2$. The lemma implies that

$$f(\delta_1)^\delta < f(\delta)^\delta < f(\delta_2)^\delta,$$ (4.7.11)

and thus

$$\{\bar{A}_x(\delta_1)\}^{\delta/\delta_1} < \bar{A}_x(\delta) < \{\bar{A}_x(\delta_2)\}^{\delta/\delta_2},$$ (4.7.12)

which allows us to estimate $\bar{A}_x(\delta)$ if the values of $\bar{A}_x(\delta_1)$ and $\bar{A}_x(\delta_2)$ are known.

For instance, let

$$\bar{A}_{50} = 0.41272 \quad \text{for} \quad i = 4\% \,,$$
$$\bar{A}_{50} = 0.34119 \quad \text{for} \quad i = 5\% \,.$$

Bounds for the net single premiums $\bar{A}_{50}$ and $\bar{a}_{50}$ for $i = 4\frac{1}{2}\%$ may now be found. From (4.7.12) with

$$\delta_1 = \ln 1.04\,, \quad \delta = \ln 1.045\,, \quad \delta_2 = \ln 1.05$$

we find immediately

$$0.37039 < \bar{A}_{50} < 0.37904\,.$$

The identity $\bar{a}_{50} = (1 - \bar{A}_{50})/\delta$ then gives

$$14.304 > \bar{a}_{50} > 14.107\,.$$

Replacing T by $K + 1$ and $\bar{a}_{\overline{t}|}$ by

$$\ddot{a}_{\overline{t}|} = \frac{1 - v^t}{d}, \quad t > 0, \tag{4.7.13}$$

we obtain the inequalities

$$\ddot{a}_x < \ddot{a}_{\overline{e_x + 1}|}, \tag{4.7.14}$$

$$A_x > v^{e_x + 1}, \tag{4.7.15}$$

$$\{\, A_x(\delta_1)\}^{\delta/\delta_1} < A_x(\delta) < \{\, A_x(\delta_2)\}^{\delta/\delta_2} \tag{4.7.16}$$

by similar arguments.

The first two derivatives of the function $\bar{A}_x(\delta)$ are

$$\begin{aligned}
\bar{A}_x'(\delta) &= -\mathrm{E}[Tv^T] = -\,(\bar{I}\bar{A})_x(\delta)\,, \\
\bar{A}_x''(\delta) &= \mathrm{E}[T^2v^T] > 0\,.
\end{aligned} \tag{4.7.17}$$

Thus $\bar{A}_x(\delta)$ is a monotonically decreasing, convex function of δ. Hence any curve segment lies below the secant,

$$\bar{A}_x(\delta) < \frac{\delta_2 - \delta}{\delta_2 - \delta_1}\, \bar{A}_x(\delta_1) + \frac{\delta - \delta_1}{\delta_2 - \delta_1}\, \bar{A}_x(\delta_2)\,, \tag{4.7.18}$$

but above the tangents

$$\begin{aligned}
\bar{A}_x(\delta) &> \bar{A}_x(\delta_1) - (\delta - \delta_1)\,(\bar{I}\bar{A})_x(\delta_1)\,, \\
\bar{A}_x(\delta) &> \bar{A}_x(\delta_2) + (\delta_2 - \delta)\,(\bar{I}\bar{A})_x(\delta_2)\,.
\end{aligned} \tag{4.7.19}$$

Sometimes one obtains narrower bounds from (4.7.18) and (4.7.19) than from (4.7.12). In the example above an improved upper bound is obtained from (4.7.18):

$$\bar{A}_{50} < 0.37687\,;$$

The lower bound for $\bar{a}_{50}$ is also improved:

$$\bar{a}_{50} > 14.157\,.$$

4.8 Payments Starting at Non-integral Ages

The initial age x will in general not be integer-valued, unless it is rounded. We shall consider calculation of $\ddot{a}_{x+u}$ for integers x and $0 < u < 1$.

Starting with the identity

$$_u p_x \,_k p_{x+u} = \,_k p_x \,_u p_{x+k} \tag{4.8.1}$$

we use *Assumption a* of Section 2.6 to find

$$(1 - u\, q_x) \,_k p_{x+u} = \,_k p_x (1 - u\, q_{x+k}) \,. \tag{4.8.2}$$

Multiplying by v^k and summing over all k we obtain

$$(1 - u\, q_x)\, \ddot{a}_{x+u} = \ddot{a}_x - u(1 + i)\, A_x \,. \tag{4.8.3}$$

Now we replace A_x by $1 - d\, \ddot{a}_x$ to obtain the desired formula:

$$\ddot{a}_{x+u} = \frac{(1 + ui)\, \ddot{a}_x - u(1 + i)}{1 - u\, q_x} \,. \tag{4.8.4}$$

By means of (4.6.1) we can rewrite the above result as

$$\ddot{a}_{x+u} = \frac{1 - u}{1 - u\, q_x}\, \ddot{a}_x + \frac{u(1 - q_x)}{1 - u\, q_x}\, \ddot{a}_{x+1} \,, \tag{4.8.5}$$

so that $\ddot{a}_{x+u}$ is a weighted mean of $\ddot{a}_x$ and $\ddot{a}_{x+1}$.

In practical applications $\ddot{a}_{x+u}$ is often approximated by linear interpolation, i.e.

$$\ddot{a}_{x+u} \approx (1 - u)\, \ddot{a}_x + u\, \ddot{a}_{x+1} \,. \tag{4.8.6}$$

The approximation is particularly good for small values of q_x, which is immediately evident from (4.8.5).

As an illustration we take $\ddot{a}_{70} = 8.0960$, $\ddot{a}_{71} = 7.7364$, $q_{70} = 0.05526$. The results are tabulated below.

u	$\ddot{a}_{70+u}$ from (4.8.4),(4.8.5)	$\ddot{a}_{70+u}$ from (4.8.6)
1/12	8.0676	8.0660
2/12	8.0389	8.0361
3/12	8.0099	8.0061
4/12	7.9806	7.9761
5/12	7.9511	7.9462
6/12	7.9213	7.9162
7/12	7.8912	7.8862
8/12	7.8609	7.8563
9/12	7.8302	7.8263
10/12	7.7992	7.7963
11/12	7.7680	7.7664

If linear interpolation is also permitted for annuities with more frequent payments,

$$\ddot{a}_{x+u}^{(m)} \approx (1-u)\,\ddot{a}_x^{(m)} + u\,\ddot{a}_{x+1}^{(m)}, \tag{4.8.7}$$

we obtain from (4.3.5) the practical approximation

$$\ddot{a}_{x+u}^{(m)} \approx \alpha(m)(1-u)\,\ddot{a}_x + \alpha(m)u\,\ddot{a}_{x+1} - \beta(m). \tag{4.8.8}$$

Similar relations may be derived for the net single premium of whole life insurances starting at a fractional age. For instance, the following is an immediate consequence of (4.8.5):

$$A_{x+u} = \frac{1-u}{1-u\,q_x}\,A_x + \frac{u(1-q_x)}{1-u\,q_x}\,A_{x+1}. \tag{4.8.9}$$

Chapter 5. Net Premiums

5.1 Introduction

An insurance policy specifies on the one hand the benefits payable by the insurer (benefits may consist of one payment or a series of payments, see Chapters 3 and 4), and on the other hand the premium(s) payable by the insured. Three forms of premium payment can be distinguished:

1. One single premium,
2. Periodic premiums of a constant amount (*level* premiums),
3. Periodic premiums of varying amounts.

For periodic premiums the duration and frequency of premium payments must be specified in addition to the premium amount(s). In principle, premiums are paid in advance.

With respect to an insurance policy, we define the *total loss* L to the insurer to be the difference between the present value of the benefits and the present value of the premium payments. This loss must be considered in the algebraic sense: an acceptable choice of the premiums must result in a range of the random variable L that includes negative as well as positive values.

A premium is called a *net premium* if it satisfies the *equivalence principle*

$$\mathrm{E}[L] = 0 \,, \tag{5.1.1}$$

i.e. if the expected value of the loss is zero. If the insurance policy is financed by a single premium, the net single premium as defined in Chapters 3 and 4 satisfies condition (5.1.1). If the premium is to be paid periodically with constant amounts, equation (5.1.1) determines the net premium uniquely. Of course, in payment mode 3 (variable premiums), equation (5.1.1) is not sufficient for the determination of the net premiums.

5.2 An Example

Let us consider a term insurance for a life of age 40 (duration: 10 years; sum insured: C, payable at the end of the year of death; premium Π payable

annually in advance while the insured is alive, but not longer than 10 years).
The loss L of the insurer is given by

$$L = \begin{cases} Cv^{K+1} - \Pi\,\ddot{a}_{\overline{K+1}|} & \text{for } K = 0, 1, \cdots, 9 \ . \\ -\Pi\,\ddot{a}_{\overline{10}|} & \text{for } K \geq 10 ; \end{cases} \tag{5.2.1}$$

here K denotes the curtate-future-lifetime of (40). The random variable L
has a discrete distribution concentrated in 11 points:

$$\begin{aligned} \Pr(L = Cv^{k+1} - \Pi\,\ddot{a}_{\overline{k+1}|}) &= {}_kp_{40}\,q_{40+k}, \quad k = 0, 1, \cdots, 9, \\ \Pr(L = -\Pi\,\ddot{a}_{\overline{10}|}) &= {}_{10}p_{40}\,. \end{aligned} \tag{5.2.2}$$

We shall determine the net annual premium. From (5.1.1) one obtains the
condition

$$C\,A^1_{40:\overline{10}|} - \Pi\,\ddot{a}_{40:\overline{10}|} = 0\,, \tag{5.2.3}$$

resulting in

$$\Pi = C\,\frac{A^1_{40:\overline{10}|}}{\ddot{a}_{40:\overline{10}|}}\,. \tag{5.2.4}$$

As an illustration, we take $i = 4\%$ and assume that the mortality of (40)
follows De Moivre's law with terminal age $\omega = 100$. This somewhat unre-
alistic assumption allows the reader to check our calculations with a pocket
calculator. We have

$$\begin{aligned} A^1_{40:\overline{10}|} &= \frac{1}{60}v + \frac{1}{60}v^2 + \cdots + \frac{1}{60}v^{10} = \frac{1}{60}\,a_{\overline{10}|} = 0.1352\,, \\ A_{40:\overline{10}|}{}^1 &= \frac{5}{6}v^{10} = 0.5630\,, \end{aligned} \tag{5.2.5}$$

so that

$$\begin{aligned} A_{40:\overline{10}|} &= 0.6982\,, \\ \ddot{a}_{40:\overline{10}|} &= (1 - A_{40:\overline{10}|})/d = 7.8476\,. \end{aligned} \tag{5.2.6}$$

(5.2.4) then gives us the net annual premium:

$$\Pi = 0.0172C\,. \tag{5.2.7}$$

The insurer cannot be expected to pay benefits in return for net premiums:
there should be a safety loading which reflects the assumed risk. In what
follows a method for determining premiums will be demonstrated, which takes
account of the incurred risk.

To this end premiums are determined by a *utility function* $u(\cdot)$: this is a
function satisfying $u'(x) > 0$ and $u''(x) < 0$, and measuring the utility that

the insurer has of a monetary amount x. More specifically, we assume that the utility function is exponential,

$$u(x) = \frac{1}{a}(1 - e^{-ax}): \tag{5.2.8}$$

the parameter $a > 0$ measures the risk aversion of the insurer. The condition (5.1.1) is now replaced by the condition

$$E[u(-L)] = u(0). \tag{5.2.9}$$

i.e. premiums should now be determined in such a way that the expected utility loss is zero. With the utility function given by (5.2.8), the annual premium must satisfy

$$E[e^{aL}] = 1. \tag{5.2.10}$$

From (5.2.2) with $_kp_{40}\, q_{40+k} = \frac{1}{60}$ and $_{10}p_{40} = 5/6$, we obtain

$$\frac{1}{60} \sum_{k=0}^{9} \exp(aCv^{k+1} - a\Pi\, \ddot{a}_{\overline{k+1}|}) + \frac{5}{6} \exp(-a\Pi\, \ddot{a}_{\overline{10}|}) = 1. \tag{5.2.11}$$

We chose $a = 10^{-6}$ arbitrarily for this example. The annual premiums obtained from (5.2.11) are tabulated below.

Sum insured C	Annual premium Π	Percent of net premium
100,000	1,790	104%
500,000	10,600	123%
1,000,000	26,400	153%
2,000,000	85,900	250%
3,000,000	221,900	430%
4,000,000	525,300	764%
5,000,000	1,073,600	1248%

Obviously, now the premium is not proportional to the sum insured, as is the case with the net premium, but increases progressively with C. This is perfectly reasonable: A sum insured of 100,000 units represents a small risk to the insurer, hence the safety loading (4%) is modest. A sum insured of 5 million, on the other hand, represents a considerable risk (at least if $a = 10^{-6}$), which, in theory, makes a safety loading of 1148% acceptable.

At first glance, this result seems to contradict insurance practice, since premiums usually are proportional to the sum insured. The contradiction can be resolved by the following consideration: Assume that the insurer charges 250% of the net premium for all values of C: then policies with a sum insured exceeding 2 million require *reinsurance*: policies with a lower sum insured are

overcharged. which compensates for the relatively high fixed costs of these
policies.

Net premiums are nevertheless of utmost importance in insurance practice.
Moreover. they are usually calculated on conservative assumptions about future interest and mortality, thus creating an implicit safety loading.

5.3 Elementary Forms of Insurance

5.3.1 Whole Life and Term Insurance

We consider a whole life insurance of 1 unit, payable at the end of the year of
death. which is to be financed by net annual premiums, which we denote by
P_x. The loss of the insurer is

$$L = v^{K+1} - P_x \ddot{a}_{\overline{K+1}|}. \tag{5.3.1}$$

From (5.1.1) it follows immediately that

$$P_x = \frac{A_x}{\ddot{a}_x}. \tag{5.3.2}$$

Representing the premium payments as the difference of two perpetuities (one
starting at time 0. the other at time $K + 1$). we obtain

$$L = \left(1 + \frac{P_x}{d}\right) v^{K+1} - \frac{P_x}{d}. \tag{5.3.3}$$

Thus

$$\text{Var}(L) = \left(1 + \frac{P_x}{d}\right)^2 \text{Var}(v^{K+1}). \tag{5.3.4}$$

This equation shows that the insurer runs a greater risk (at least expressed by
the variance of L) if the insurance is financed by net annual premiums rather
than by a net single premium.

Equation (5.3.2) can be used to derive two formulae for P_x which can be
given instructive interpretations. Dividing equation (4.2.8) by $\ddot{a}_x$ we obtain
the identity

$$\frac{1}{\ddot{a}_x} = d + P_x. \tag{5.3.5}$$

This identity has the following interpretation: A debt of 1 can be amortised
by annual advance payments of $1/\ddot{a}_x$. Alternatively one can pay advance
interest (d) on the debt each year. and the amount of 1 at time $K + 1$: the
net annual premium for the corresponding life insurance is P_x. The identity
(5.3.5) means that the the total annual payments are the same in either way.

The identity (5.3.5) reminds us of another identity from the theory of interest,

$$\frac{1}{\ddot{a}_{\overline{n}|}} = d + \frac{1}{\ddot{s}_{\overline{n}|}}, \tag{5.3.6}$$

which also has a similar interpretation (see Section 1.8).

Replacing $\ddot{a}_x$ by $(1 - A_x)/d$ in (5.3.2), we find

$$P_x = \frac{d\,A_x}{1 - A_x}. \tag{5.3.7}$$

The equivalent identity

$$P_x = d\,A_x + P_x\,A_x \tag{5.3.8}$$

may be interpreted as follows: A coverage of 1 unit can be financed by annual payments of P_x; on the other hand, one can imagine that an amount of A_x is borrowed to pay the net single premium. Interest on the debt of A_x is paid annually in advance, and the debt is repaid at the end of the year of death; the annual premium for the corresponding life insurance is $P_x\,A_x$. The identity (5.3.8) shows that the total annual payments are the same either way.

We shall consider a term insurance of duration n (sum insured 1 unit, payable at the end of the year of death). The net annual premium is denoted by $P^1_{x:\overline{n}|}$. The insurer's loss is

$$L = \begin{cases} v^{K+1} - P^1_{x:\overline{n}|}\,\ddot{a}_{\overline{K+1}|} & \text{for } K = 0, 1, \cdots, n-1, \\ -P^1_{x:\overline{n}|}\,\ddot{a}_{\overline{n}|} & \text{for } K \geq n, \end{cases} \tag{5.3.9}$$

or, as in (5.3.3),

$$L = -P^1_{x:\overline{n}|}\,\ddot{a}_{\overline{n}|} + (1 + P^1_{x:\overline{n}|}\,\ddot{a}_{\overline{n-K-1}|})v^{K+1}I_{\{K<n\}}. \tag{5.3.10}$$

The net annual premium is, of course,

$$P^1_{x:\overline{n}|} = \frac{A^1_{x:\overline{n}|}}{\ddot{a}_{x:\overline{n}|}}. \tag{5.3.11}$$

5.3.2 Pure Endowments

Let the sum insured be 1 unit and the duration n. The net annual premium is denoted by $P_{x:\overline{n}|}^{1}$. The loss of the insurer is

$$L = \begin{cases} -P_{x:\overline{n}|}^{1}\,\ddot{a}_{\overline{K+1}|} & \text{for } K = 0, 1, \cdots, n-1 \\ v^n - P_{x:\overline{n}|}^{1}\,\ddot{a}_{\overline{n}|} & \text{for } K \geq n. \end{cases} \tag{5.3.12}$$

The net annual premium is obviously

$$P_{x:\overline{n}|}^{1} = \frac{A_{x:\overline{n}|}^{1}}{\ddot{a}_{x:\overline{n}|}}. \tag{5.3.13}$$

5.3.3 Endowments

The net annual premium is denoted by $P_{x:\overline{n}|}$. The equations

$$P_{x:\overline{n}|} = \frac{A_{x:\overline{n}|}}{\ddot{a}_{x:\overline{n}|}} \tag{5.3.14}$$

and

$$P_{x:\overline{n}|} = P^1_{x:\overline{n}|} + P_{x:\overline{n}|}^{\phantom{x:\overline{n}|}1} \tag{5.3.15}$$

are obvious. The insurer's loss is the sum of (5.3.9) and (5.3.12).

In analogy with (5.3.5) and (5.3.8) we have

$$\frac{1}{\ddot{a}_{x:\overline{n}|}} = d + P_{x:\overline{n}|}, \tag{5.3.16}$$

$$P_{x:\overline{n}|} = d\, A_{x:\overline{n}|} + P_{x:\overline{n}|}\, A_{x:\overline{n}|}, \tag{5.3.17}$$

with the corresponding interpretations. Equation (5.3.17) can also be obtained by adding the relations

$$P^1_{x:\overline{n}|} = d\, A^1_{x:\overline{n}|} + P_{x:\overline{n}|}\, A^1_{x:\overline{n}|}, \tag{5.3.18}$$

$$P_{x:\overline{n}|}^{\phantom{x:\overline{n}|}1} = d\, A_{x:\overline{n}|}^{\phantom{x:\overline{n}|}1} + P_{x:\overline{n}|}\, A_{x:\overline{n}|}^{\phantom{x:\overline{n}|}1}, \tag{5.3.19}$$

each of these having an interpretation similar to that of (5.3.8).

5.3.4 Deferred Life Annuities

The net annual premium payable during the deferment period for a life annuity-due of 1 p.a. starting at time n, is $P_{x:\overline{n}|}^{\phantom{x:\overline{n}|}1}\ddot{a}_{x+n}$.

5.4 Premiums Paid m Times a Year

If the net annual premium is paid by m installments of equal size, the superscript "(m)" is is attached to the appropriate premium symbol. The net annual premiums

$$P_x^{(m)},\ P_{x:\overline{n}|}^{(m)},\ P^1_{x:\overline{n}|}{}^{(m)},\ P_{x:\overline{n}|}^{\phantom{x:\overline{n}|}1\,(m)}$$

are obtained by replacing $\ddot{a}_x$, resp. $\ddot{a}_{x:\overline{n}|}$, by $\ddot{a}_x^{(m)}$, resp. $\ddot{a}_{x:\overline{n}|}^{(m)}$, in the denominators of (5.3.2), (5.3.11), (5.3.13), (5.3.14). The net annual premium of an endowment paying 1 unit is for instance

$$P_{x:\overline{n}|}^{(m)} = A_{x:\overline{n}|} / \ddot{a}_{x:\overline{n}|}^{(m)}. \tag{5.4.1}$$

The expression may be readily evaluated by means of formula (4.3.9).

In order to compare $P_{x:\overline{n}|}^{(m)}$ with $P_{x:\overline{n}|}$, we substitute in (5.4.1)

$$A_{x:\overline{n}|} = P_{x:\overline{n}|} \ddot{a}_{x:\overline{n}|}, \tag{5.4.2}$$

$$\ddot{a}_{x:\overline{n}|}^{(m)} = \frac{d}{d^{(m)}} \ddot{a}_{x:\overline{n}|} - \beta(m) A_{x:\overline{n}|}^1 \tag{5.4.3}$$

and obtain

$$P_{x:\overline{n}|}^{(m)} = \frac{P_{x:\overline{n}|}}{d/d^{(m)} - \beta(m) P_{x:\overline{n}|}^1}. \tag{5.4.4}$$

If we now write the last result in the form

$$P_{x:\overline{n}|} = \ddot{a}_{\overline{1}|}^{(m)} P_{x:\overline{n}|}^{(m)} - \beta(m) P_{x:\overline{n}|}^{(m)} P_{x:\overline{n}|}^1, \tag{5.4.5}$$

two reasons for the relation $P_{x:\overline{n}|} < P_{x:\overline{n}|}^{(m)}$ become apparent.

Analogous relations hold for other insurances, e.g.

$$P_x = \ddot{a}_{\overline{1}|}^{(m)} P_x^{(m)} - \beta(m) P_x^{(m)} P_x, \tag{5.4.6}$$

$$P_{x:\overline{n}|}^1 = \ddot{a}_{\overline{1}|}^{(m)} P_{x:\overline{n}|}^{1\ (m)} - \beta(m) P_{x:\overline{n}|}^{1\ (m)} P_{x:\overline{n}|}^1, \tag{5.4.7}$$

$$P_{x:\overline{n}|}^{\ 1} = \ddot{a}_{\overline{1}|}^{(m)} P_{x:\overline{n}|}^{\ 1\ (m)} - \beta(m) P_{x:\overline{n}|}^{\ 1\ (m)} P_{x:\overline{n}|}^1. \tag{5.4.8}$$

Equation (5.4.6) is the limit of (5.4.5) as $n \to \infty$. Equation (5.4.5) is the sum of equations (5.4.7) and (5.4.8).

5.5 A General Type of Life Insurance

We return to the general type of life insurance introduced in Section 3.4. Let c_j be the sum insured in the jth year after policy issue. We assume that the insurance is to be financed by annual premiums $\Pi_0, \Pi_1, \Pi_2, \cdots, \Pi_k$ being the premium due at time k. The insurer's loss is

$$L = c_{K+1} v^{K+1} - \sum_{k=0}^{K} \Pi_k v^k. \tag{5.5.1}$$

The premiums are net premiums if they satisfy the equation

$$\sum_{k=0}^{\infty} c_{k+1} v^{k+1} \ _k p_x \ q_{x+k} = \sum_{k=0}^{\infty} \Pi_k v^k \ _k p_x. \tag{5.5.2}$$

The model is more general than it may appear at first glance. If negative values are permitted for the Π_k, it includes pure endowments and life annuities. For instance, the endowment of Section 5.3.3 is obtained by setting

$$c_1 = c_2 = \cdots = c_n = 1, \qquad\qquad c_{n+1} = c_{n+2} = \cdots = 0,$$

$$\Pi_0 = \Pi_1 = \cdots = \Pi_{n-1} = P_{x:\overline{n}|}, \quad \Pi_n = -1, \quad \Pi_{n+1} = \Pi_{n+2} = \cdots = 0. \tag{5.5.3}$$

5.6 Policies with Premium Refund

A large variety of insurance forms and payment plans occur in practical insurance. This makes it impractical to derive the net single premium explicitly for every possible combination. The fundamental rule to be followed in a given situation is to specify the insurer's loss L, and then to apply the condition (5.1.1). This procedure will be illustrated with an example.

A pure endowment with 1 unit payable after n years is issued with the provision that, in case of death before n, the premiums paid will be refunded without interest. What should the net annual premium be if the premium charged is to exceed the net annual premium by 40%? (The 40% loading is used to cover expenses).

We let P denote the net annual premium. The insurer's loss is obviously

$$
L = \begin{cases} (K+1)(1.4P)v^{K+1} - P\,\ddot{a}_{\overline{K+1}|} & \text{for } K = 0, 1, \cdots, n-1 \,, \\ v^n - P\,\ddot{a}_{\overline{n}|} & \text{for } K \geq n. \end{cases}
\tag{5.6.1}
$$

The expected loss is

$$
1.4\,P\,(IA)^1_{x:\overline{n}|} + A_{x:\overline{n}|}^{\;\;1} - P\,\ddot{a}_{x:\overline{n}|},
\tag{5.6.2}
$$

and application of (5.1.1) leads to the premium

$$
P = \frac{A_{x:\overline{n}|}^{\;\;1}}{\ddot{a}_{x:\overline{n}|} - 1.4\,(IA)^1_{x:\overline{n}|}}.
\tag{5.6.3}
$$

5.7 Stochastic Interest

The interest rate that will apply in future years is of course not known. Thus it seems reasonable to ask why future interest rates have not been modelled as a stochastic process. Two reasons have led us to refrain from such a model: 1) Life insurance is particularly concerned with the *long term* development of interest rates and no commonly accepted stochastic model exists for making long term predictions. 2) A reasonable assumption is that the remaining lifetimes of the insured lives are, essentially, independent random variables. With a fixed interest assumption, the insurer's losses from different policies become independent random variables. The probability distribution of the aggregate loss can then simply be obtained by convolution. In particular, the variance of the aggregate loss is the sum of the individual variances, which facilitates the use of the normal approximation. Stochastic independence between policies would be lost with the introduction of a stochastic interest rate, since all policies are affected by the same interest development.

Thus we shall continue using the assumption of a fixed interest rate. The practical evaluation of an insurance cover should analyse different interest scenarios. It is also possible to let the interest assumption vary over time, say using i_j as the interest assumption for year j. This would not lead to mathematical complications, but would make the notation more laborious, so that we shall not follow in this direction.

Chapter 6. Net Premium Reserves

6.1 Introduction

Consider an insurance policy which is financed by net premiums. At the time of policy issue, the expected present value of future premiums equals the expected present value of future benefit payments, making the expected loss L of the insurer zero.

This equivalence between future payments and future benefits does not, in general, exist at a later time. Thus we define a random variable $_tL$ as the difference at time t between the present value of future benefit payments and the present value of future premium payments; we assume that $_tL$ is not identically equal to zero, and we also assume that $T > t$. The *net premium reserve* at time t is denoted by $_tV$, and it is defined as the conditional expectation of $_tL$, given that $T > t$.

Life insurance policies are usually designed in such a way that the net premium reserve is positive, or at least non-negative, for the insured should at all times have an interest in continuing the insurance. Thus the expected value of future benefits will always exceed the expected value of future premium payments. To compensate for this liability the insurer should always reserve sufficient funds to cover the difference of these expected values, i.e. the net premium reserve $_tV$.

6.2 Two Examples

The net premium reserve at the end of the kth policy year for an endowment insurance (duration: n, sum insured: 1 payable after n years or at the end of the year of death, annual premiums) is denoted by $_kV_{x:\overline{n}|}$ and given by the expression

$$_kV_{x:\overline{n}|} = A_{x+k:\overline{n-k}|} - P_{x:\overline{n}|}\ddot{a}_{x+k:\overline{n-k}|}, \quad k = 0, 1, \cdots, n-1. \qquad (6.2.1)$$

Obviously $_0V_{x:\overline{n}|} = 0$ because of the definition of net premiums.

The net premium reserve at the end of year k of the corresponding term insurance is denoted by $_kV^1_{x:\overline{n}|}$. It is given by

$$_kV^1_{x:\overline{n}|} = A^{\ 1}_{x+k:\overline{n-k}|} - P^1_{x:\overline{n}|}\ddot{a}_{x+k:\overline{n-k}|}. \qquad (6.2.2)$$

For a numerical illustration, we assume a sum insured of 1000 units, initial age $x = 40$, and the duration $n = 10$. The net premium reserve is thus $1000\,_kV_{40:\overline{10}|}$ ($1000\,_kV^1_{40:\overline{10}|}$) for $k = 0, 1, \cdots, 9$. As in Section 5.2 we assume $i = 4\%$ and use De Moivre's survival function with $\omega = 100$ for our calculations.

As a first step we find the net annual premium 88.96 for the endowment and 17.225 for the term insurance. The development of the net premium reserves is tabulated below; the entries can easily be verified with a pocket calculator. Though De Moivre's law is not very realistic, the net premium reserves follow a characteristic pattern.

Development of net premium reserve for an endowment and a term insurance

k	$\ddot{a}_{40+k:\overline{10-k}|}$	$A_{40+k:\overline{10-k}|}$ $\times 1000$	$_kV_{40:\overline{10}|}$ $\times 1000$	$A^{\ 1}_{40+k:\overline{10-k}|}$ $\times 1000$	$_kV^1_{40:\overline{10}|}$ $\times 1000$
0	7.84805	698.15	0	135.18	0.0
1	7.24269	721.44	77	126.02	1.3
2	6.60433	745.99	158	116.08	2.3
3	5.93076	771.89	244	105.30	3.1
4	5.21956	799.25	335	93.61	3.7
5	4.46813	828.15	431	80.94	4.0
6	3.67365	858.71	532	67.22	3.9
7	2.83306	891.04	639	52.36	3.6
8	1.94305	925.27	752	36.27	2.8
9	1.00000	961.54	873	18.85	1.6

The net premium reserve of the endowment grows steadily and approaches the sum insured towards the end. The net premium reserve of 872.58 at the end of the 9th year can be easily verified: The sum of this net premium reserve and the last premium payment of 88.96, plus interest on both, must be sufficient to cover the payment of 1000 one year later.

The net premium reserve of the term insurance is very small and nearly constant. Initially it grows since the premium slightly exceeds that of a corresponding one-year term insurance. Towards the end the net premium reserve decreases again since the insurer has no obligation if the insured survives. The sum of the net premium reserve at the end of the 9th year (1.62) and the last premium (17.23) is exactly sufficient to cover a one-year term insurance for a 49-year old (18.85).

6.3 Recursive Considerations

We return to the general life insurance introduced in Section 5.5. The net premium reserve at the end of year k is, according to the definition,

$$_kV = \sum_{j=0}^{\infty} c_{k+j+1} v^{j+1} {}_j p_{x+k} \, q_{x+k+j} - \sum_{j=0}^{\infty} \Pi_{k+j} v^j {}_j p_{x+k} \,. \qquad (6.3.1)$$

In order to derive a relation between $_kV$ and $_{k+h}V$, we substitute

$$_j p_{x+k} = {}_h p_{x+k} \, {}_{j-h} p_{x+k+h} \qquad (6.3.2)$$

in all except the first h terms of (6.3.1), and use $j' = j - h$ as summation index. The resulting relation between $_kV$ and $_{k+h}V$ is

$$_kV + \sum_{j=0}^{h-1} \Pi_{k+j} v^j {}_j p_{x+k} = \sum_{j=0}^{h-1} c_{j+k+1} v^{j+1} {}_j p_{x+k} \, q_{x+k+j} + {}_h p_{x+k} v^h {}_{k+h}V \,. \qquad (6.3.3)$$

It is not surprising that this relation has the following interpretation: If the insured is alive at the end of year k, then the net premium reserve, together with the expected present value of the premiums to be paid during the next h years is just sufficient to pay for the life insurance during those years, plus a pure endowment of $_{k+h}V$ at the end of year $k + h$.

A recursive equation for the net premium reserve is obtained by letting $h = 1$:

$$_kV + \Pi_k = v[c_{k+1} \, q_{x+k} + {}_{k+1}V \, p_{x+k}] \,. \qquad (6.3.4)$$

Thus the net premium reserve may be calculated recursively in two directions: 1) One may calculate $_1V, {}_2V, \cdots$ successively, starting with the initial value $_0V = 0$. 2) If the insurance is of finite duration n, then one may calculate $_{n-1}V, {}_{n-2}V, \cdots$ in this order, starting with the known value of $_nV$. For example, in the numerical example of Section 6.2 we have $_{10}V = 1000$ for the endowment, and $_{10}V = 0$ for the term insurance.

Equation (6.3.4) shows that the sum of the net premium reserve at time k and the premium equals the expected present value of the funds needed at the end of the year (these being c_{k+1} in case of death, else $_{k+1}V$). Another interpretation becomes evident when one writes

$$_kV + \Pi_k = v[{}_{k+1}V + (c_{k+1} - {}_{k+1}V) \, q_{x+k}] \,. \qquad (6.3.5)$$

The amount of $_{k+1}V$ is needed in any case. The additional amount needed if the insured dies , $c_{k+1} - {}_{k+1}V$, is the *net amount at risk*.

Equation (6.3.5) shows that the premium can be decomposed into two components, $\Pi_k = \Pi_k^s + \Pi_k^r$, where

$$\Pi_k^s = {}_{k+1}V v - {}_kV \qquad (6.3.6)$$

is the *savings premium* used to increase the net premium reserve, and

$$\Pi_k^r = (c_{k+1} - {}_{k+1}V)v\, q_{x+k} \tag{6.3.7}$$

is the premium of a one-year term insurance to cover the net amount at risk, or *risk premium*. Thus the operation in year $k + 1$ may be interpreted as a combination of a pure savings operation and a one-year term insurance. We are assuming, of course, that the insured is alive at time k.

Multiplying (6.3.6) by $(1+i)^{j-k}$ and summing over $k = 0, 1, \cdots, j - 1$, we obtain

$$_jV = \sum_{k=0}^{j-1}(1+i)^{j-k}\, \Pi_k^s, \tag{6.3.8}$$

which shows that the net premium reserve is the accumulated value of the savings premiums paid since policy issue.

The decomposition into savings premium and risk premium in the numerical example of Section 6.2 is tabulated below.

Decomposition into savings premium and risk premium

k	Endowment		Term insurance	
	Π_k^s	Π_k^r	Π_k^s	Π_k^r
0	74.17	14.79	1.22	16.00
1	75.24	13.71	0.97	16.26
2	76.43	12.53	0.70	16.53
3	77.74	11.22	0.42	16.81
4	79.18	9.78	0.12	17.10
5	80.77	8.18	−0.19	17.41
6	82.53	6.43	−0.52	17.74
7	84.47	4.49	−0.87	18.09
8	86.60	2.36	−1.24	18.46
9	88.96	0.00	−1.62	18.85

Writing (6.3.5) as

$$\Pi_k + d\,{}_{k+1}V = ({}_{k+1}V - {}_kV) + \Pi_k^r, \tag{6.3.9}$$

we see that the premium, plus the interest earned on the net premium reserve, serves to modify (increase or decrease) the net premium reserve and to finance the risk premium. This equation is apparently a generalisation of (3.6.7).

Multiplying (6.3.5) by $(1 + i)$, we obtain an equation similar to (6.3.9):

$$\Pi_k + i({}_kV + \Pi_k) = ({}_{k+1}V - {}_kV) + (c_{k+1} - {}_{k+1}V)\, q_{x+k}. \tag{6.3.10}$$

Equations (6.3.9) and (6.3.10) differ in that the valuation is performed at time k in (6.3.9), but at time $k + 1$ in (6.3.10).

6.4 The Survival Risk

The derivations of the previous section are valid also if $c_{k+1} < {}_{k+1}V$, i.e. if the net amount at risk is negative. But in this case the analysis may also be modified. We start by expressing (6.3.4) as

$${}_kV + \Pi_k = c_{k+1}v + ({}_{k+1}V - c_{k+1})v\,p_{x+k} \,. \tag{6.4.1}$$

The amount of c_{k+1} is needed in any case; in case of survival, an additional amount of ${}_{k+1}V - c_{k+1}$ falls due. The financial transactions during year $k+1$ may thus be allocated partly to pure savings, and partly to a pure endowment with a face amount of ${}_{k+1}V - c_{k+1}$. The premium Π_k may be viewed as the sum of a modified savings premium,

$$\hat{\Pi}_k^s = c_{k+1}v - {}_kV \,, \tag{6.4.2}$$

and the *survival risk premium*

$$\hat{\Pi}_k^r = ({}_{k+1}V - c_{k+1})v\,p_{x+k} \,. \tag{6.4.3}$$

We note that the savings component will often be negative, too. Equation (6.4.1) may also be expressed as

$$\Pi_k + dc_{k+1} = (c_{k+1} - {}_kV) + \hat{\Pi}_k^r \,, \tag{6.4.4}$$

a formula which reminds us of (6.3.9).

The decomposition of premium into (6.4.2) and (6.4.3) is not very common, and in what follows we shall not use it.

6.5 The Net Premium Reserve of a Whole Life Insurance

Consider the whole life insurance introduced in Section 5.3.1. Its net premium reserve at the end of year k is denoted by ${}_kV_x$ and is by definition

$${}_kV_x = A_{x+k} - P_x\,\ddot{a}_{x+k} \,. \tag{6.5.1}$$

We shall derive some equivalent formulae.

Replacing A_{x+k} by $1 - d\,\ddot{a}_{x+k}$, we find

$${}_kV_x = 1 - (P_x + d)\,\ddot{a}_{x+k} \,. \tag{6.5.2}$$

Now, replacing $P_x + d$ by $1/\ddot{a}_x$, we obtain

$${}_kV_x = 1 - \frac{\ddot{a}_{x+k}}{\ddot{a}_x} \,. \tag{6.5.3}$$

The formula

$$_kV_x = \frac{A_{x+k} - A_x}{1 - A_x} \tag{6.5.4}$$

may be verified if we replace $\ddot{a}_x$ by $(1 - A_x)/d$ and $\ddot{a}_{x+k}$ by $(1 - A_{x+k})/d$. The identity $P_{x+k}\, \ddot{a}_{x+k} = A_{x+k}$ with (6.5.1) gives

$$_kV_x = \left(1 - \frac{P_x}{P_{x+k}}\right) A_{x+k} \tag{6.5.5}$$

and

$$_kV_x = (P_{x+k} - P_x)\, \ddot{a}_{x+k}. \tag{6.5.6}$$

Finally we replace $\ddot{a}_{x+k}$ by $1/(P_{x+k} + d)$ to find

$$_kV_x = \frac{P_{x+k} - P_x}{P_{x+k} + d}. \tag{6.5.7}$$

The multitude of different formulae may seem confusing. Apart from (6.5.1) the formulae (6.5.2), (6.5.5) and (6.5.6) are important because they are easily interpreted and because they may be generalised to other types of insurance.

Formula (6.5.2) expresses the fact that the net premium reserve equals the sum insured, less the expected present value of future premiums and unused interest. This reminds us of the identity $A_x = 1 - d\,\ddot{a}_x$, which has a similar interpretation.

Equation (6.5.5) may be interpreted by recognising that the future premiums of P_x may serve to finance a whole life insurance with face amount P_x/P_{x+k}; the net premium reserve is then used to finance the remaining face amount of $1 - P_x/P_{x+k}$.

If the whole life insurance were to be bought at age $x + k$ the net annual premium would be P_{x+k}. The *premium difference formula* (6.5.6) shows that the net premium reserve is the expected present value of the shortfall of the premiums.

Equations (6.5.3), (6.5.4) and (6.5.7) are of lesser importance and have no obvious interpretation. However, they allow generalisation to endowment insurance.

6.6 Net Premium Reserves at Fractional Durations

We return to the general insurance discussed in Section 6.3. Let us assume that the insured is alive at time $k+u$ (k an integer, $0 < u < 1$), and denote the net premium reserve by $_{k+u}V$. Similarly to (6.3.5), the net premium reserve can be expressed by

$$_{k+u}V = {}_{k+1}V v^{1-u} + (c_{k+1} - {}_{k+1}V)v^{1-u}\,_{1-u}q_{x+k+u}. \tag{6.6.1}$$

Assumption a of Section 2.6 implies

$$_{1-u}q_{x+k+u} = \frac{(1-u)\,q_{x+k}}{1-u\,q_{x+k}}, \tag{6.6.2}$$

which permits direct evaluation of $_{k+u}V$.

We can also express $_{k+u}V$ in terms of $_kV$. In order to do so we substitute (6.6.2) in (6.6.1) and use (6.3.7) and (6.3.6). We obtain

$$_{k+u}V = (_kV + \Pi_k^s)(1+i)^u + \frac{1-u}{1-u\,q_{x+k}}\,\Pi_k^r(1+i)^u. \tag{6.6.3}$$

In Section 6.3 we saw that the operation in year $k+1$ could be decomposed; equation (6.6.3) gives the corresponding decomposition at a fractional duration: The first term is the balance of a fictitious savings account at time $k+u$, and the second term is the part of the risk premium which is still "unearned" at time $k+u$.

A third possible formula is

$$_{k+u}V = \frac{1-u}{1-u\,q_{x+k}}(_kV + \Pi_k)(1+i)^u + \left\{1 - \frac{1-u}{1-u\,q_{x+k}}\right\}\,_{k+1}V\,v^{1-u}. \tag{6.6.4}$$

This shows that $_{k+u}V$ is a weighted average of the accumulated value of $(_kV + \Pi_k)$ and the discounted value of $_{k+1}V$; the weights are identical to the weights in (4.8.5), for $k = 0$. To prove (6.6.4), we replace Π_k by $\Pi_k^s + \Pi_k^r$; definition (6.3.6) then shows that (6.6.4) is equivalent to (6.6.3).

In practical applications an approximation based on linear interpolation is often used:

$$_{k+u}V \approx (1-u)(_kV + \Pi_k) + u\,_{k+1}V. \tag{6.6.5}$$

To see how good this approximation is, we replace Π_k by $\Pi_k^s + \Pi_k^r$ and $_{k+1}V$ by $(_kV + \Pi_k^s)(1+i)$. The approximation is then

$$_{k+u}V \approx (_kV + \Pi_k^s)(1+ui) + (1-u)\,\Pi_k^r, \tag{6.6.6}$$

which permits direct comparison with (6.6.3)

6.7 Allocation of the Overall Loss to Policy Years

We continue the discussion of the general life insurance. For $k = 0, 1, \cdots$, we define Λ_k to be the loss incurred by the insurer during the year $k+1$; thus the beginning of the year is used as reference point on the time scale. Three cases can be distinguished: 1) The insured has died before time k, 2) the insured dies during year $k+1$, 3) the insured survives to $k+1$. The random variable Λ_k is thus defined by

$$\Lambda_k = \begin{cases} 0 & \text{if } K \le k-1, \\ c_{k+1}v - (_kV + \Pi_k) & \text{if } K = k, \\ _{k+1}Vv - (_kV + \Pi_k) & \text{if } K \ge k+1. \end{cases} \tag{6.7.1}$$

Replacing Π_k by $\Pi_k^s + \Pi_k^r$ and using (6.3.6), we find

$$\Lambda_k = \begin{cases} 0 & \text{if } K \le k - 1 \,, \\ -\Pi_k^r + (c_{k+1} - {}_{k+1}V)v & \text{if } K = k \,, \\ -\Pi_k^r & \text{if } K \ge k + 1 \,. \end{cases} \tag{6.7.2}$$

Thus, if the insured is alive at time k, Λ_k is the loss produced by the one-year term insurance covering the net amount at risk.

The overall loss of the insurer is given by equation (5.5.1). The obvious result

$$L = \sum_{k=0}^{\infty} \Lambda_k v^k \tag{6.7.3}$$

may be verified directly through (6.7.1). Of course the sum is finite, running from 0 to K.

Using (6.7.2) and (6.3.7) we find

$$E[\Lambda_k | K \ge k] = 0 \,, \tag{6.7.4}$$

which again implies

$$E[\Lambda_k] = E[\Lambda_k | K \ge k] \Pr(K \ge k) = 0 \,. \tag{6.7.5}$$

While (6.7.3) is generally valid, the validity of (6.7.5) requires that each year's payments are offset against the *net* premium reserve of that year.

The classical *Hattendorff's Theorem* states that

$$\mathrm{Cov}(\Lambda_k, \Lambda_j) = 0 \quad \text{for } k \ne j \,, \tag{6.7.6}$$

$$\mathrm{Var}(L) = \sum_{k=0}^{\infty} v^{2k} \mathrm{Var}(\Lambda_k) \,. \tag{6.7.7}$$

The second formula states that the variance of the insurer's overall loss can be allocated to individual policy years, and it is a direct consequence of the first formula and (6.7.3). The first formula is not directly evident since the random variables $\Lambda_0, \Lambda_1, \cdots$ are not independent.

In a proof of (6.7.6) we may assume $k < j$ without loss of generality. Then one has

$$\begin{aligned} \mathrm{Cov}(\Lambda_k, \Lambda_j) &= E[\Lambda_k \cdot \Lambda_j] \\ &= E[\Lambda_k \cdot \Lambda_j | K \ge j] \, \Pr(K \ge j) \\ &= -\Pi_k^r E[\Lambda_j | K \ge j] \Pr(K \ge j) \\ &= 0 \,; \end{aligned} \tag{6.7.8}$$

here (6.7.4) has been used in the last step.

The variance of Λ_k may be calculated as follows:

$$
\begin{aligned}
\mathrm{Var}(\Lambda_k) &= \mathrm{E}[\Lambda_k^2] \\
&= \mathrm{E}[\Lambda_k^2 | K \geq k]\, \mathrm{Pr}(K \geq k) \\
&= \mathrm{Var}(\Lambda_k | K \geq k)\, \mathrm{Pr}(K \geq k) \\
&= (c_{k+1} - {}_{k+1}V)^2 v^2\, p_{x+k}\, q_{x+k}\, \mathrm{Pr}(K \geq k) \\
&= (c_{k+1} - {}_{k+1}V)^2 v^2\, {}_{k+1}p_x\, q_{x+k}\, .
\end{aligned}
\tag{6.7.9}
$$

Substituting this into (6.7.7) we finally find

$$
\mathrm{Var}(L) = \sum_{k=0}^{\infty} v^{2k+2}(c_{k+1} - {}_{k+1}V)^2\, {}_{k+1}p_x\, q_{x+k}\, .
\tag{6.7.10}
$$

We now assume that the insured is alive at time h (h an integer), and consider the loss defined in Section 6.1, being the difference between the expected present values of future benefit payments and future premium payments. In analogy to (6.7.10) we have

$$
\mathrm{Var}({}_hL) = \sum_{k=0}^{\infty} v^{2k+2}(c_{h+k+1} - {}_{h+k+1}V)^2\, {}_{k+1}p_{x+h}\, q_{x+h+k}\, .
\tag{6.7.11}
$$

To prove this we consider a hypothetical insurance, issued at age $x + h$ and financed by the "premiums"

$$
\tilde{\Pi}_0 = \Pi_h + {}_hV\,, \quad \tilde{\Pi}_k = \Pi_{h+k} \ \text{ for } k = 1, 2, \cdots\, .
\tag{6.7.12}
$$

The variance of L may be easily evaluated by means of equation (6.7.10). The results, for the numerical example of Section 6.2, have been compiled in the table below.

Calculation of the variance of L by policy years

k	Endowment	Term insurance
0	12905	15114
1	9918	13940
2	7393	12864
3	5292	11876
4	3584	10970
5	2240	10140
6	1231	9379
7	535	8682
8	131	8043
9	0	7457
Sum	43229	108465

We see that the variance of L is much smaller for the endowment ($43\,229$) than for the term insurance ($108\,465$).

Equation (6.7.10) is useful in evaluating the influence of the financing method on the variance of L, when the benefit plan is fixed. Consider for instance a pure endowment, with $c_1 = c_2 = \cdots = 0$. The variance of L increases with the net premium reserve. Thus financing by a net single premium leads to a greater variance than financing by net annual premiums.

6.8 Conversion of an Insurance

In a technical sense the net premium reserve "belongs" to the insured and may in principle be used to help finance a modification of the insurance policy at any time.

A classical example is the conversion of an insurance policy into a paid-up insurance, i.e. one for which no further premium payments are required. Consider a whole life insurance issued at age x with a sum insured of 1 unit, and financed by annual premiums of P_x. Assume that the insured is alive at time k, but, for whatever reasons, unable to pay further premiums. In such a situation the net premium reserve of $_kV_x$ could be considered as the net single premium for a whole life insurance with a sum insured of

$$_kV_x / A_{x+k} = 1 - P_x / P_{x+k}\,, \tag{6.8.1}$$

see (6.5.5). Such conversions into paid-up insurance with reduced benefits are very common for endowments (for which the net premium reserve is substantial).

A type of insurance known as *"universal life"* or *"flexible life"*, made possible by modern data processing, offers the insured a maximum degree of flexibility. Here the insured may adjust the parameters of the insurance periodically (e.g. annually). The insured who "owns" the premium reserve of $_kV$ at time k, may change any two of the following parameters:

- Π_k, the next premium to be paid,
- c_{k+1}, the sum insured in case of death during the next year,
- $_{k+1}V$, the target value of his "savings" one year ahead.

The third parameter is then determined by the recursive formula (6.3.4). In other words, the insured effectively decides next year's premium, as well as its decomposition into savings premium and risk premium. Certain restrictions are usually imposed to reduce the risk of antiselection; for instance, the new sum insured (c_{k+1}) should not exceed the former sum insured (c_k) by more than a predetermined percentage, which could, possibly, depend on the inflation rate.

6.9 Technical Gain

Consider the general life insurance of Section 6.3, and let us assume that the insured is alive at time k. We assume further that the actually earned interest rate during year $k+1$ is i'. The *technical gain* at the end of the year is then

$$G_{k+1} = \begin{cases} ({}_kV + \Pi_k)(1 + i') - c_{k+1} & \text{if } K = k\ , \\ ({}_kV + \Pi_k)(1 + i') - {}_{k+1}V & \text{if } K \geq k+1\ . \end{cases} \qquad (6.9.1)$$

Essentially there are two ways in which this technical gain can be decomposed:

Method 1

Replacing $1 + i'$ by $(i' - i) + (1 + i)$ in (6.9.1), one obtains

$$G_{k+1} = ({}_kV + \Pi_k)(i' - i) - \Lambda_k(1 + i)\,. \qquad (6.9.2)$$

The technical gain thus consists of an *investment gain* and a *mortality gain*.

Method 2

Since the operation during year $k+1$ may be considered as part savings and part insurance, a reasonable approach is to allocate the technical gain accordingly:

$$G_{k+1} = G_{k+1}^s + G_{k+1}^r\,. \qquad (6.9.3)$$

Here

$$G_{k+1}^s = ({}_kV + \Pi_k^s)(i' - i) \qquad (6.9.4)$$

is the gain from savings, and

$$G_{k+1}^r = \begin{cases} \Pi_k^r(1 + i') - (c_{k+1} - {}_{k+1}V) & \text{if } K = k\ , \\ \Pi_k^r(1 + i') & \text{if } K \geq k+1 \end{cases} \qquad (6.9.5)$$

is the gain from the insurance. The latter may again be decomposed into

$$G_{k+1}^r = \Pi_k^r(i' - i) - \Lambda_k(1 + i)\,, \qquad (6.9.6)$$

see (6.7.2). The last equation shows the connection to *Method 1*.

When the technical interest rate i is chosen conservatively, the technical gain, respectively G_{k+1}^s, will usually be positive. If this gain is to be passed on to the insured through increased benefits, then *Method 2* is preferable, since the gain from savings may be written as

$$G_{k+1}^s = {}_{k+1}V\,v(i' - i)\,. \qquad (6.9.7)$$

The future benefits may then be increased uniformly by

$$v(i' - i)100\%\,, \qquad (6.9.8)$$

provided that the insured agrees to future premiums being increased by the same factor. As a result of this profit sharing, the insured will obtain a modified insurance policy for which

$$\tilde{c}_{k+1+h} = v(1 + i')c_{k+1+h}, \quad \tilde{\Pi}_{k+h} = v(1 + i')\,\Pi_{k+h} \qquad (6.9.9)$$

for $h = 0, 1, \cdots$. This will be the case if the insured is alive at the end of the year. In case of death ($K = k$), the gain from savings G^s_{k+1} may be paid in addition to the sum insured of c_{k+1}.

6.10 Procedure for Pure Endowments

Consider a pure endowment ($c_1 = c_2 = \cdots = 0$). The technical gain at the end of year $k + 1$ is

$$G_{k+1} = \begin{cases} (_kV + \Pi_k)(1 + i') & \text{if } K = k\ , \\ (_kV + \Pi_k)(1 + i') - _{k+1}V & \text{if } K \geq k + 1. \end{cases} \qquad (6.10.1)$$

Since it is desirable to have an investment gain only in the case of survival ($K \geq k + 1$), we decompose the technical gain in a slightly different way:

$$G_{k+1} = G^I_{k+1} + G^{II}_{k+1}, \qquad (6.10.2)$$

with

$$G^I_{k+1} = \begin{cases} 0 & \text{if } K = k\ , \\ _{k+1}V v(i' - i) & \text{if } K \geq k + 1\ , \end{cases} \qquad (6.10.3)$$

and

$$G^{II}_{k+1} = \begin{cases} p_{x+k}\,_{k+1}V v(1 + i') & \text{if } K = k\ , \\ -\,q_{x+k}\,_{k+1}V v(1 + i') & \text{if } K \geq k + 1\ . \end{cases} \qquad (6.10.4)$$

The proof of this decomposition follows from (6.10.1) and the fact that

$$(_kV + \Pi_k) = p_{x+k}\,_{k+1}V v, \qquad (6.10.5)$$

see (6.3.4). Note that the expectation of G^{II}_{k+1} is zero, which is not the case with the expectation of G^r_{k+1}.

If the insured survives, the gain given by (6.10.3) may be used to increase the benefits, provided future premiums are increased accordingly, by a factor determined by (6.9.8).

Similar derivations to the above may be made for *life annuities*, simply by equating r_k, the contractually agreed payment at time k, to $-\Pi_k$. For instance, if a pension fund has an investment yield of i' during a year, the interest gained from the annuities may be used to increase all annuities by the factor given in (6.9.8).

6.11 The Continuous Model

Let us finally consider the continuous counterpart to the general life insurance of Section 6.3.

The insurance is now determined by two functions, the amount insured $c(t)$ and the premium rate $\Pi(t)$, both at the moment t, $t \geq 0$. The net premium reserve at time t is

$$V(t) = \int_0^\infty c(t+h)v^h \, {}_hp_{x+t} \, \mu_{x+t+h} dh - \int_0^\infty \Pi(t+h)v^h \, {}_hp_{x+t} dh \,. \quad (6.11.1)$$

The premium rate can be decomposed into a savings component,

$$\Pi^s(t) = V'(t) - \delta V(t) \,, \quad (6.11.2)$$

and a risk component,

$$\Pi^r(t) = (c(t) - V(t))\mu_{x+t} \,. \quad (6.11.3)$$

That $\Pi(t)$ is the sum of those two components establishes *Thiele's Differential Equation*:

$$\Pi(t) + \delta V(t) = V'(t) + \Pi^r(t) \,; \quad (6.11.4)$$

it is the continuous version of (6.3.9) and (6.3.10) and has a similar interpretation.

In the special case that

$$c(t) = 1 \,, \quad \Pi(t) = 0 \,, \quad V(t) = \bar{A}_{x+t} \,, \quad (6.11.5)$$

equation (6.11.4) leads to (3.6.11). If

$$c(t) = 0 \,, \quad \Pi(t) = -1 \,, \quad V(t) = \bar{a}_{x+t} \,, \quad (6.11.6)$$

equation (6.11.4) confirms (4.6.6).

Working within the continuous model simplifies matters. There is for instance only one method for analysing the technical gain, instead of two, as in the discrete model of Sections 6.9 and 6.10.

We assume that the insured is alive at time t, and that the actual force of interest at time t is $\delta(t)$. The technical gain in the infinitesimal time interval from t to $t+dt$, which we denote be $G(t, t+dt)$, can be decomposed into

$$G(t, t+dt) = G^s(t, t+dt) + G^r(t, t+dt) \,; \quad (6.11.7)$$

here

$$G^s(t, t+dt) = (\delta(t) - \delta)V(t)dt \quad (6.11.8)$$

is the investment gain, and

$$G^r(t, t+dt) = \begin{cases} -(c(t) - V(t)) & \text{if } t < T < t+dt \,, \\ \Pi^r(t)dt & \text{if } T > t+dt \,, \end{cases} \quad (6.11.9)$$

is the mortality gain. Note that the probability of death is $\mu_{x+t}dt$, and the probability of survival is $1 - \mu_{x+t}dt$, so that the expected value of $G^r(t, t+dt)$ is zero. Note also that

$$\text{Var}[G^r(t, t+dt)|T > t] = (c(t) - V(t))^2 \mu_{x+t}dt, \tag{6.11.10}$$

$$\text{Var}[G^r(t, t+dt)] = (c(t) - V(t))^2 \, {}_tp_x\mu_{x+t}dt, \tag{6.11.11}$$

and

$$\begin{aligned}
\text{Var}(L) &= \int_0^\infty v^{2t}\text{Var}[G^r(t, t+dt)] \\
&= \int_0^\infty v^{2t}(c(t) - V(t))^2 \, {}_tp_x\mu_{x+t}dt, \tag{6.11.12}
\end{aligned}$$

in analogy with (6.7.7) and (6.7.10).

Using a life annuity as an example, we shall demonstrate how the investment gain may be used to increase the benefits continuously. Assume that a continuous life annuity with constant payment rate $r(t)$ is guaranteed at time t. The net premium reserve at time t is thus

$$V(t) = r(t)\,\bar{a}_{x+t}. \tag{6.11.13}$$

At time $t + dt$ the payment rate is to be increased to $r(t+dt) = r(t) + r'(t)dt$, the cost of which must be covered by the investment gain. This leads to the condition

$$G^s(t, t+dt) = r'(t)\,dt\,\bar{a}_{x+t}. \tag{6.11.14}$$

Using (6.11.8) and (6.11.13) we obtain a differential equation for $r(t)$, viz.

$$(\delta(t) - \delta)r(t) = r'(t), \tag{6.11.15}$$

with solution

$$r(t) = r(0)\exp\left\{\int_0^t (\delta(s) - \delta)ds\right\}, \tag{6.11.16}$$

which is in accordance with the result derived at the end of Section 6.10.

We have seen in this and the last two sections how the investment gain can be used to increase the benefits on an individually equitable basis. On the other hand, it is impossible to pass on the mortality gain to the insured on an individual basis: Death of the insured causes a mortality loss (in case of life insurance) or a mortality gain (in case of an annuity), which naturally cannot be passed on to the insured.

It is, however, possible to pass on mortality gain (or loss) to a group of insureds. This will be demonstrated by an example which is reminiscent of the historical *Tontines*.

Consider a group consisting initially of n persons; all have the same initial age x and are initially guaranteed a life annuity of constant rate 1. It has been agreed to pass on any mortality gain (or loss) to the annuitants in the form of increased (or decreased) future payments. What will be the value of $r_k(t)$, the annuity rate at time t, if then only k of the initially n persons are still alive?

Assuming that k persons are alive at time t and that all survive to time $t + dt$, the mortality gain will be negative; per survivor it amounts to

$$G^r(t, t + dt) = \Pi^r(t)dt = -r_k(t)\,\bar{a}_{x+t}\mu_{x+t}dt\,, \qquad (6.11.17)$$

see (6.11.9). The reduction in the annuity rate then follows from the condition

$$G^r(t, t + dt) = r'_k(t)dt\,\bar{a}_{x+t}\,, \qquad (6.11.18)$$

which, in turn, implies the differential equation

$$r'_k(t) = -r_k(t)\mu_{x+t}\,. \qquad (6.11.19)$$

If one of the k persons dies at time t, an immediate mortality gain of $r_k(t)\,\bar{a}_{x+t}$ results; this is distributed among the $k - 1$ survivors to increase the annuity rate. The new annuity rates follow from the condition that the net premium reserve should be unchanged:

$$k\,r_k(t)\,\bar{a}_{x+t} = (k - 1)r_{k-1}(t)\,\bar{a}_{x+t}\,. \qquad (6.11.20)$$

Thus one may write

$$r_{k-1}(t) = \frac{k}{k - 1}r_k(t)\,, \quad k = 2, 3, \cdots, n\,. \qquad (6.11.21)$$

The explicit solution is found using (6.11.19), (6.11.21) and the initial condition $r_n(0) = 1$ to be

$$r_k(t) = \frac{n}{k}\,{}_tp_x\,, \quad k = 1, 2, \cdots, n\,. \qquad (6.11.22)$$

Is is easy to check and not at all surprising that the organiser of such an arrangement may in fact be considered to be functioning purely as a banker as long as at least one person lives, and finally to be making a profit of

$$r_1(z)\,\bar{a}_{x+z} = n\,{}_zp_x\,\bar{a}_{x+z}\,, \qquad (6.11.23)$$

if z denotes the time of the last person's death.

Chapter 7. Multiple Decrements

7.1 The Model

In this chapter we extend the model introduced in Chapter 2 and reinterpret the remaining lifetime random variable T.

Assume that the person under consideration is in a specific status at age x. The person leaves that status at time T due to one of m mutually exclusive causes of decrement (numbered conveniently from 1 to m). We shall study a pair of random variables, the remaining lifetime in the specified status T and the *cause of decrement J*.

In a classical example, disability insurance, the initial status is *"Active"*, and possible causes of decrement are *"Disablement"* and *"Death"*.

In another setting T is the remaining lifetime of (x), distinguishing between two causes of decrement, death by *"Accident"* and by *"Other causes"*. This model is appropriate in connection with insurances which provide double indemnity on accidental death.

The joint probability distribution of T and J can be written in terms of the density functions $g_1(t), \cdots, g_m(t)$, so that

$$g_j(t)dt = \Pr(t < T < t + dt, J = j) \tag{7.1.1}$$

is the probability of decrement by cause j in the infinitesimal time interval $(t, t + dt)$. Obviously

$$g(t) = g_1(t) + \cdots + g_m(t) . \tag{7.1.2}$$

If the decrement occurs at time t, the conditional probability of j being the cause of decrement is

$$\Pr(J = j | T = t) = \frac{g_j(t)}{g(t)} . \tag{7.1.3}$$

We introduce the symbols

$$_t q_{j,x} = \Pr(T < t, J = j) \tag{7.1.4}$$

or, more generally,

$$_t q_{j,x+s} = \Pr(T < s + t, J = j | T > s) . \tag{7.1.5}$$

The latter probability is calculated as follows:

$$t q_{j,x+s} = \int_s^{s+t} g_j(z)dz/[1 - G(s)]. \tag{7.1.6}$$

7.2 Forces of Decrement

For a life (x) the force of decrement at age $x + t$ in respect of the cause j is defined by

$$\mu_{j,x+t} = \frac{g_j(t)}{1 - G(t)} = \frac{g_j(t)}{t p_x}. \tag{7.2.1}$$

The aggregate force of decrement is

$$\mu_{x+t} = \mu_{1,x+t} + \cdots + \mu_{m,x+t}, \tag{7.2.2}$$

see (7.1.2) and definition (2.2.1).

Equation (7.1.1) can be expressed as

$$\Pr(t < T < t + dt, J = j) = {}_t p_x \mu_{j,x+t} dt. \tag{7.2.3}$$

Furthermore,

$$\Pr(J = j | T = t) = \frac{\mu_{j,x+t}}{\mu_{x+t}}. \tag{7.2.4}$$

If all forces of decrement are known, the joint distribution of T and J may be determined by first using (7.2.2) and (2.2.6) to determine ${}_t p_x$ and then determining $g_j(t)$ from (7.2.1).

7.3 The Curtate Lifetime of (x)

If the one-year probabilities of decrement,

$$q_{j,x+k} = \Pr(T < k + 1, J = j | T > k) \tag{7.3.1}$$

are known for $k = 0, 1, \cdots$ and $j = 1, \cdots, m$, the joint probability distribution of the curtate time $K = [T]$ and the cause of decrement J may be evaluated. Start by observing that

$$q_{x+k} = q_{1,x+k} + \cdots + q_{m,x+k}, \tag{7.3.2}$$

from which ${}_k p_x$ can be calculated; then

$$\Pr(K = k, J = j) = {}_k p_x \, q_{j,x+k} \tag{7.3.3}$$

for $k = 0, 1, \cdots$ and $j = 1, \cdots, m$.

The joint distribution of T and J can be computed under suitable assumptions concerning probabilities of decrement at fractional ages. A popular assumption is that $_uq_{j,x+k}$ is a linear function of u for $0 < u < 1$, k an integer, i.e.

$$_uq_{j,x+k} = u\, q_{j,x+k} \,. \tag{7.3.4}$$

This assumption implies *Assumption a* of Section 2.6, which may be verified by summation over all j. From (7.3.4) follows

$$g_j(k+u) = {}_kp_x\, q_{j,x+k}\,; \tag{7.3.5}$$

together with the identity $_{k+u}p_x = {}_kp_x(1 - u\, q_{x+k})$ this yields

$$\mu_{j,x+k+u} = \frac{q_{j,x+k}}{1 - u\, q_{x+k}}\,. \tag{7.3.6}$$

Assumption (7.3.4) has the obvious advantage known from Chapter 2, that K and S become independent random variables, and that S will have a uniform distribution between 0 and 1. In addition one has

$$\Pr(J = j | K = k, S = u) = \frac{q_{j,x+k}}{q_{x+k}}\,, \tag{7.3.7}$$

a consequence of (7.2.4) and (7.3.6). The last relation states that the conditional probability of decrement by cause j is constant during the year. In closing we summarise that S has a uniform distribution between 0 and 1, independently of the pair (K, J), and that the distribution of (K, J) is given by (7.3.3).

7.4 A General Type of Insurance

Consider an insurance which provides for payment of the amount $c_{j,k+1}$ at the end of year $k+1$, if decrement by cause j occurs during that year. The present value of the insured benefit is thus

$$Z = c_{J,K+1}v^{K+1}\,, \tag{7.4.1}$$

and the net single premium is

$$\mathrm{E}(Z) = \sum_{j=1}^{m}\sum_{k=0}^{\infty} c_{j,k+1}v^{k+1}\, {}_kp_x\, q_{j,x+k}\,. \tag{7.4.2}$$

If the insurance provides for payment immediately on death, the present value of the insured benefit is

$$Z = c_J(T)v^T\,, \tag{7.4.3}$$

and the net single premium is

$$\mathrm{E}(Z) = \sum_{j=1}^{m} \int_{0}^{\infty} c_j(t) v^t g_j(t) dt \, . \tag{7.4.4}$$

This expression may be evaluated numerically by splitting each of the m integrals, viz.

$$\mathrm{E}(Z) = \sum_{j=1}^{m} \sum_{k=0}^{\infty} \int_{0}^{1} c_j(k+u) v^{k+u} g_j(k+u) du \, . \tag{7.4.5}$$

Use of assumption (7.3.4) allows us to substitute (7.3.5) in the expression above. Thus (7.4.5) assumes the form (7.4.2) if we write

$$c_{j,k+1} = \int_{0}^{1} c_j(k+u)(1+i)^{1-u} du \, . \tag{7.4.6}$$

In a practical calculation the approximation

$$c_{j,k+1} \approx c_j(k + \frac{1}{2})(1+i)^{\frac{1}{2}} \tag{7.4.7}$$

will often be sufficiently accurate. The above derivations show that the evaluation of the net single premium in the continuous model (7.4.3) can be reduced to a calculation within the discrete model (7.4.1).

The insured's exit from the initial status will not always result in a single payment; another possibility is the initiation of a life annuity. If, for instance, the cause $j = 1$ denotes disablement, then $c_1(t)$ could be the net single premium of a temporary life annuity starting at age $x + t$. Thus in the general model the "payments" $c_{j,k+1}$ (respectively $c_j(t)$) may themselves be expected values; however, the formulae (7.4.2) and (7.4.4) remain valid.

7.5 The Net Premium Reserve

Let us assume that the general insurance benefits of Section 7.4 are supported by annual premiums of $\Pi_0, \Pi_1, \Pi_2, \cdots$. The net premium reserve at the end of year k is then

$$_kV = \sum_{j=1}^{m} \sum_{h=0}^{\infty} c_{j,k+h+1} v^{h+1} {}_h p_{x+k} \, q_{j,x+k+h} - \sum_{h=0}^{\infty} \Pi_{k+h} v^{h} {}_h p_{x+k} \, . \tag{7.5.1}$$

The recursive equation

$$_kV + \Pi_k = {}_{k+1}V v \, p_{x+k} + \sum_{j=1}^{m} c_{j,k+1} v \, q_{j,x+k} \tag{7.5.2}$$

is a generalisation of (6.3.4). It may be expressed as

$$_kV + \Pi_k = {}_{k+1}Vv + \sum_{j=1}^{m}(c_{j,k+1} - {}_{k+1}V)v\, q_{j,x+k} \, . \qquad (7.5.3)$$

Thus the premium may again be decomposed into two components, the *savings premium*

$$\Pi_k^s = {}_{k+1}Vv - {}_kV \qquad (7.5.4)$$

to increment the net premium reserve, and the *risk premium*

$$\Pi_k^r = \sum_{j=1}^{m}(c_{j,k+1} - {}_{k+1}V)v\, q_{j,x+k} \qquad (7.5.5)$$

to insure the net amount at risk for one year.

The insurer's overall loss

$$L = c_{J,K+1}v^{K+1} - \sum_{k=0}^{K}\Pi_k v^k \qquad (7.5.6)$$

may again be decomposed into

$$L = \sum_{k=0}^{\infty}\Lambda_k v^k \, , \qquad (7.5.7)$$

where

$$\Lambda_k = \begin{cases} 0 & \text{if } K \leq k - 1 \, , \\ -\Pi_k^r + (c_{J,k+1} - {}_{k+1}V)v & \text{if } K = k \, , \\ -\Pi_k^r & \text{if } K \geq k + 1 \, , \end{cases} \qquad (7.5.8)$$

is the insurer's loss in year $k + 1$, evaluated at time k. Hattendorff's Theorem (Equations (6.7.4)–(6.7.7)) remains valid. The variance of L is most conveniently evaluated by the formula

$$\text{Var}(L) = \sum_{k=0}^{\infty}\text{Var}(\Lambda_k|K \geq k)v^{2k}\,{}_kp_x \, , \qquad (7.5.9)$$

now with

$$\text{Var}(\Lambda_k|K \geq k) = \sum_{j=1}^{m}(c_{j,k+1} - {}_{k+1}V)^2 v^2\, q_{j,x+k} - (\Pi_k^r)^2 \, . \qquad (7.5.10)$$

The verification of the last formula is left to the reader.

The activities in year $k+1$ thus may be regarded as a combination of pure savings on the one hand, and a one-year insurance transaction on the other hand. The latter can be decomposed into m elementary coverages, one for each cause of decrement. We may interpret the premium component

$$\Pi_{j,k}^r = (c_{j,k+1} - {}_{k+1}V)v\, q_{j,x+k} \qquad (7.5.11)$$

as paying for a one-year insurance of the amount $(c_{j,k+1} - {}_{k+1}V)$, which covers the risk from decrement cause j. The insurer's loss during year $k+1$ may be decomposed accordingly:

$$\Lambda_k = \Lambda_{1,k} + \Lambda_{2,k} + \cdots + \Lambda_{m,k}, \tag{7.5.12}$$

if we define

$$\Lambda_{j,k} = \begin{cases} 0 & \text{if } K \le k-1, \\ -\Pi_{j,k}^r + (c_{j,k+1} - {}_{k+1}V)v & \text{if } K = k \text{ and } J = j, \\ -\Pi_{j,k}^r & \text{if } K = k \text{ and } J \ne j, \text{ or } K \ge k+1. \end{cases} \tag{7.5.13}$$

The technical gain at the end of the year,

$$G_{k+1} = \begin{cases} ({}_kV + \Pi_k)(1+i') - c_{J,k+1} & \text{if } K = k, \\ ({}_kV + \Pi_k)(1+i') - {}_{k+1}V & \text{if } K \ge k+1, \end{cases} \tag{7.5.14}$$

may similarly be decomposed into $m+1$ components. For instance, the decomposition method 1 (Section 6.9) leads to

$$G_{k+1} = ({}_kV + \Pi_k)(i' - i) - \sum_{j=1}^{m} \Lambda_{j,k}(1+i). \tag{7.5.15}$$

7.6 The Continuous Model

The model of Section 6.11 can be generalised to the multiple decrement model of this chapter. Assume that the insured benefit is defined by (7.4.3) and that premium is paid continuously, with $\Pi(t)$ denoting the premium rate at time t. The overall loss of the insurer is thus

$$L = c_J(T)v^T - \int_0^T \Pi(t)v^t dt. \tag{7.6.1}$$

The net premium reserve at time t is given by

$$V(t) = \sum_{j=1}^{m} \int_0^\infty c_j(t+h)v^h \, {}_hp_{x+t}\mu_{j,x+t+h}dh - \int_0^\infty \Pi(t+h)v^h \, {}_hp_{x+t}dh. \tag{7.6.2}$$

The premium rate $\Pi(t)$ can be decomposed into a savings component $\Pi^s(t)$, see (6.11.2), and a risk component

$$\Pi^r(t) = \sum_{j=1}^{m}(c_j(t) - V(t))\mu_{j,x+t}; \tag{7.6.3}$$

Thiele's differential equation (6.11.4) remains valid.

The technical gain derived from the insurance component in the infinitesimal interval from t to $t + dt$ is denoted by $G^r(t, t + dt)$. It is obvious that

$$G^r(t, t + dt) = \begin{cases} 0 & \text{if } T < t \text{ ,} \\ -(c_J(t) - V(t)) & \text{if } t < T < t + dt \text{ ,} \\ \Pi^r(t)dt & \text{if } T > t + dt \text{ .} \end{cases} \qquad (7.6.4)$$

As a consequence we have

$$\begin{aligned} \text{Var}[G^r(t, +dt)|T > t] &= \text{E}[\{G^r(t, t + dt)\}^2|T > t] \\ &= \sum_{j=1}^{m}(c_j(t) - V(t))^2 \mu_{j,x+t}dt \end{aligned} \qquad (7.6.5)$$

and

$$\text{Var}[G^r(t, t + dt)] = \sum_{j=1}^{m}(c_j(t) - V(t))^2 \, {}_tp_x\mu_{j,x+t}dt . \qquad (7.6.6)$$

Finally one obtains

$$\begin{aligned} \text{Var}(L) &= \int_0^\infty v^{2t}\text{Var}[G^r(t, t + dt)] \\ &= \sum_{j=1}^{m}\int_0^\infty v^{2t}(c_j(t) - V(t))^2 \, {}_tp_x\mu_{j,x+t}dt . \end{aligned} \qquad (7.6.7)$$

Note that this result is simpler than its discrete counterpart, see (7.5.9) and (7.5.10); this is not surprising in view of (7.5.10): the risk premium for the infinitesimal interval is $\Pi^r(t)\, dt$, so its square vanishes in the limit. From (7.6.7) it is also evident that the variance of L may be decomposed by causes of decrement.

Chapter 8. Multiple Life Insurance

8.1 Introduction

Consider m lives with initial ages $x_1, x_2, \cdots, x_m$. For simplicity we denote the future lifetime of the kth life, $T(x_k)$ in the notation of Chapter 2, by T_k $(k = 1, \cdots, m)$. On the basis of these m elements we shall define a status u with a future lifetime $T(u)$. We shall accordingly denote by ${}_tp_u$ the conditional probability that the status u is still intact at time t, given that the status existed at time 0; the symbols q_u, μ_{u+t} etc., are defined in a similar way. We shall also consider annuities which are defined in terms of u. The symbol $\ddot{a}_u$, for instance, denotes the net single premium of an annuity-due with 1 unit payable annually, as long as u remains intact. We shall also analyse insurances with a benefit payable at the failure of the status u. The symbol $\bar{A}_u$ would for instance denote the net single premium of an insured benefit of 1 unit, payable immediately upon the failure of u.

8.2 The Joint-Life Status

The status
$$u = x_1 : x_2 : \cdots : x_m \tag{8.2.1}$$
is defined to exist as long as all m participating lives survive. The failure time of this *joint-life status* is
$$T(u) = \text{Minimum}(T_1, T_2, \cdots, T_m). \tag{8.2.2}$$

We shall assume in what follows that the random variables $T_1, T_2, \cdots, T_m$ are independent. The probability distribution of the failure time of status (8.2.1) is then given by

$$
\begin{aligned}
{}_tp_{x_1:x_2:\cdots:x_m} &= \Pr(T(u) > t) \\
&= \Pr(T_1 > t, T_2 > t, \cdots, T_m > t) \\
&= \prod_{k=1}^{m} \Pr(T_k > t) = \prod_{k=1}^{m} {}_tp_{x_k}.
\end{aligned} \tag{8.2.3}
$$

The instantaneous failure rate of the joint-life status is, according to (2.2.5):

$$\mu_{u+t} = -\frac{d}{dt}\ln\ {}_tp_u = -\frac{d}{dt}\sum_{k=1}^{m}\ln\ {}_tp_{x_k} = \sum_{k=1}^{m}\mu_{x_k+t}\,. \tag{8.2.4}$$

This identity is reminiscent of (7.2.2). Note, however, that unlike the identity in Chapter 7, the identity (8.2.4) presupposes that $T_1,\cdots,T_m$ are independent.

The principles of Chapters 3 and 4 may now be applied to calculate, for example, the net single premium for an insurance payable on the first death,

$$A_{x_1:x_2:\cdots:x_m} = \sum_{k=0}^{\infty} v^{k+1}\ {}_kp_{x_1:x_2:\cdots:x_m}\ q_{x_1+k:x_2+k:\cdots:x_m+k}\,. \tag{8.2.5}$$

The net single premium for a joint-life annuity-due is

$$\ddot{a}_{x_1:x_2:\cdots:x_m} = \sum_{k=0}^{\infty} v^{k}\ {}_kp_{x_1:x_2:\cdots:x_m}\,. \tag{8.2.6}$$

Identities similar to those derived in Chapter 4 will be valid, for example

$$1 = d\,\ddot{a}_{x_1:x_2:\cdots:x_m} + A_{x_1:x_2:\cdots:x_m}\,. \tag{8.2.7}$$

The definitions and derivations of Chapters 5 and 6 can be generalised by replacing (x) by (u).

If we denote by $\overline{n}|$ the status which fails at time n, i.e.

$$T(\overline{n}|) = n\,, \tag{8.2.8}$$

then $T(x:\overline{n}|) = \mathrm{Minimum}(T(x),n)$; it is then evident that the net single premium symbols $\bar{A}_{x:\overline{n}|}$ (endowment) and $\bar{a}_{x:\overline{n}|}$ (temporary annuity) are in accordance with the joint-life notation.

8.3 Simplifications

A significant simplification results if all lives are subject to the same Gompertz mortality law, i.e.

$$\mu_{x_k+t} = Bc^{x_k+t}\,,\quad t\geq 0,\quad k = 1,\cdots,m\,. \tag{8.3.1}$$

After solving the equation

$$c^{x_1} + c^{x_2} + \cdots + c^{x_m} = c^{w} \tag{8.3.2}$$

for w, the instantaneous joint-life failure rate may be expressed by

$$\mu_{u+t} = \mu_{w+t}\,,\quad t\geq 0\,. \tag{8.3.3}$$

This implies that the failure rate of the joint-life status follows the same Gompertz mortality law as an individual life with "initial age" w. All calculations in respect of the joint-life status may then be performed in terms of the single life (w). As an example we have

$$A_{x_1:x_2:\cdots:x_m} = A_w,\qquad (8.3.4)$$

and

$$a_{x_1:x_2:\cdots:x_m} = a_w.\qquad (8.3.5)$$

Some simplification also results if all lives follow the same Makeham mortality law,

$$\mu_{x_k+t} = A + Bc^{x_k+t}.\qquad (8.3.6)$$

Let w be the solution of the equation

$$c^{x_1} + c^{x_2} + \cdots + c^{x_m} = mc^w,\qquad (8.3.7)$$

then (8.2.4) implies that

$$\mu_{u+t} = m\mu_{w+t} = \mu_{w+t:w+t:\cdots:w+t},\quad t \geq 0.\qquad (8.3.8)$$

This means that the m lives aged $x_1, x_2, \cdots, x_m$ may be replaced by m lives of the same "initial age" w. As an example,

$$a_{x_1:x_2:\cdots:x_m} = a_{w:w:\cdots:w}.\qquad (8.3.9)$$

Note that the age w defined by (8.3.7) is a sort of mean of the component ages $x_1, x_2, \cdots, x_m$, while the age w defined by (8.3.2) exceeds all component ages $x_1, x_2, \cdots, x_m$.

The simplifications presented in this section, albeit very elegant, have lost much of their practical value. Nowadays formulae like (8.2.3), (8.2.5) or (8.2.6) may be evaluated directly.

8.4 The Last-Survivor Status

The *last-survivor status*

$$u = \overline{x_1 : x_2 : \cdots : x_m}\qquad (8.4.1)$$

is defined to be intact while at least one of the m lives survives, so that it fails with the last death:

$$T(u) = \text{Maximum}(T_1, T_2, \cdots, T_m).\qquad (8.4.2)$$

The joint-life status and the last-survivor status may be visualized by electric circuits: The status (8.2.1) corresponds to connection in series of the m components, while the status (8.4.1) corresponds to a parallel connection.

Probabilities and net single premiums in respect of a last-survivor status may be calculated using certain joint-life statuses. To see this, the reader should recall the inclusion-exclusion formula in probability theory. Letting $B_1, B_2, \cdots, B_m$ denote events, the probability of their union is

$$\Pr(B_1 \cup B_2 \cup \cdots \cup B_m) = S_1 - S_2 + S_3 - \cdots + (-1)^{m-1} S_m ; \qquad (8.4.3)$$

here S_k denotes the symmetric sum

$$S_k = \sum \Pr(B_{j_1} \cap B_{j_2} \cap \cdots \cap B_{j_k}), \qquad (8.4.4)$$

where the summation ranges over all $\binom{m}{k}$ subsets of k events.

Denoting by B_k the event that the kth life still lives at time t, we obtain from (8.4.3)

$$_t p_{\overline{x_1 : x_2 : \cdots : x_m}} = S_1^t - S_2^t + S_3^t - \cdots + (-1)^{m-1} S_m^t, \qquad (8.4.5)$$

with the notation

$$S_k^t = \sum {}_t p_{x_{j_1} : x_{j_2} : \cdots : x_{j_k}}. \qquad (8.4.6)$$

Multiplying equation (8.4.5) by v^t and summing over t, we obtain an analogous formula for the net single premium of a last-survivor annuity:

$$\ddot{a}_{\overline{x_1 : x_2 : \cdots : x_m}} = S_1^{\ddot{a}} - S_2^{\ddot{a}} + S_3^{\ddot{a}} - \cdots + (-1)^{m-1} S_m^{\ddot{a}} ; \qquad (8.4.7)$$

here we have defined

$$S_k^{\ddot{a}} = \sum \ddot{a}_{x_{j_1} : x_{j_2} : \cdots : x_{j_k}}. \qquad (8.4.8)$$

Consider now an insured benefit of 1, payable upon the last death. Its net single premium may be calculated as follows:

$$\begin{aligned} A_{\overline{x_1 : x_2 : \cdots : x_m}} &= 1 - d\, \ddot{a}_{\overline{x_1 : x_2 : \cdots : x_m}} \\ &= 1 - d(S_1^{\ddot{a}} - S_2^{\ddot{a}} + S_3^{\ddot{a}} - \cdots). \end{aligned} \qquad (8.4.9)$$

Let us define the symmetric sums

$$S_k^A = \sum A_{x_{j_1} : x_{j_2} : \cdots : x_{j_k}}. \qquad (8.4.10)$$

Substituting

$$S_k^{\ddot{a}} = \frac{\binom{m}{k} - S_k^A}{d} \qquad (8.4.11)$$

in (8.4.9), we obtain the formula

$$A_{\overline{x_1 : x_2 : \cdots : x_m}} = S_1^A - S_2^A + S_3^A - \cdots + (-1)^{m-1} S_m^A. \qquad (8.4.12)$$

Note the similarity of equations (8.4.5), (8.4.7) and (8.4.12). Similar formulae may be derived for the net single premium of fractional or continuous annuities, or insurances payable immediately on the last death.

As an illustration, consider the case of 3 lives with initial ages x, y and z. In this case we have, for instance,

$$\ddot{a}_{\overline{x\,:\,y\,:\,z}} = S_1^{\ddot{a}} - S_2^{\ddot{a}} + S_3^{\ddot{a}} , \qquad (8.4.13)$$

with

$$
\begin{aligned}
S_1^{\ddot{a}} &= \ddot{a}_x + \ddot{a}_y + \ddot{a}_z , \\
S_2^{\ddot{a}} &= \ddot{a}_{x:y} + \ddot{a}_{x:z} + \ddot{a}_{y:z} , \\
S_3^{\ddot{a}} &= \ddot{a}_{x:y:z} .
\end{aligned}
\qquad (8.4.14)
$$

The net single premiums $\ddot{a}_{x:y}$, $\ddot{a}_{x:z}$, $\ddot{a}_{y:z}$, as well as $\ddot{a}_{x:y:z}$ may be calculated using equations (8.2.3) and (8.2.6).

8.5 The General Symmetric Status

We define the status

$$u = \frac{k}{\overline{x_1 : x_2 : \cdots : x_m}} \qquad (8.5.1)$$

to last as long as at least k of the initial m lives survive, i.e. it fails upon the $(m-k+1)$th death. The joint-life status $(k = m)$ and the last-survivor status $(k = 1)$ are obviously special cases of this status.

The status

$$u = \frac{[k]}{\overline{x_1 : x_2 : \cdots : x_m}} \qquad (8.5.2)$$

is defined to be intact when exactly k of the m lives survive. The status starts to exist at the $(m-k)$th death and fails at the $(m-k+1)$th death. The status (8.5.2) may be of interest in the context of annuities, but not for insurances.

A general solution follows from the *Schuette-Nesbitt formula*, which is the topic of the next section. For arbitrarily chosen coefficients $c_0, c_1, \cdots, c_m$ one has

$$\sum_{k=0}^{m} c_k \, {}_t p_{\overline{x_1 : x_2 : \cdots : x_m}}^{[k]} = \sum_{j=0}^{m} \Delta^j c_0 S_j^t \qquad (8.5.3)$$

and, similarly,

$$\sum_{k=0}^{m} c_k \, \ddot{a}_{\overline{x_1 : x_2 : \cdots : x_m}}^{[k]} = \sum_{j=0}^{m} \Delta^j c_0 S_j^{\ddot{a}} . \qquad (8.5.4)$$

Here the values S_j^t and $S_j^{\ddot{a}}$ are defined by (8.4.6) and (8.4.8), for $j = 1, 2, \cdots, m$; we also define $S_0^t = 1$ and $S_0^{\ddot{a}} = \ddot{a}_{\overline{\infty}}$.

For arbitrarily chosen coefficients $d_1, d_2, \cdots, d_m$ one also has

$$\sum_{k=1}^{m} d_k \, {}_tp_{\overline{x_1 \,:\, x_2 \,:\, \cdots \,:\, x_m}}^{\quad\quad\quad\quad\quad k} = \sum_{j=1}^{m} \Delta^{j-1} d_1 S_j^t \qquad (8.5.5)$$

and, similarly,

$$\sum_{k=1}^{m} d_k \, \ddot{a}_{\overline{x_1 \,:\, x_2 \,:\, \cdots \,:\, x_m}}^{\quad\quad\quad\quad\quad k} = \sum_{j=1}^{m} \Delta^{j-1} d_1 S_j^{\ddot{a}} . \qquad (8.5.6)$$

The last two formulae are a consequence of the former two: with

$$c_0 = 0, \quad c_k = d_1 + \cdots + d_k , \qquad (8.5.7)$$

the left hand sides of (8.5.5) and (8.5.6) assume the form of (8.5.3) and (8.5.4).

The expressions (8.5.5) and (8.5.6) have the advantage that they can be generalized to life insurances:

$$\sum_{k=1}^{m} d_k \, A_{\overline{x_1 \,:\, x_2 \,:\, \cdots \,:\, x_m}}^{\quad\quad\quad\quad\quad k} = \sum_{j=1}^{m} \Delta^{j-1} d_1 S_j^{A} . \qquad (8.5.8)$$

This equation is obtained from (8.5.6) in the same way as (8.4.12) was obtained from (8.4.7).

As an illustration we consider a continuous annuity payable to 4 lives of initial ages w, x, y, z. The payment rate starts at 8 and is reduced by 50% for each death. The net single premium of this annuity is obviously

$$8\,\bar{a}_{\overline{w \,:\, x \,:\, y \,:\, z}}^{\quad\quad\quad [4]} + 4\,\bar{a}_{\overline{w \,:\, x \,:\, y \,:\, z}}^{\quad\quad\quad [3]} + 2\,\bar{a}_{\overline{w \,:\, x \,:\, y \,:\, z}}^{\quad\quad\quad [2]} + \bar{a}_{\overline{w \,:\, x \,:\, y \,:\, z}}^{\quad\quad\quad [1]} \; ; \quad (8.5.9)$$

thus we have the coefficients $c_0 = 0$, $c_1 = 1$, $c_2 = 2$, $c_3 = 4$, $c_4 = 8$. The difference table is as follows:

k	c_k	Δc_k	$\Delta^2 c_k$	$\Delta^3 c_k$	$\Delta^4 c_k$
0	0	1	0	1	0
1	1	1	1	1	
2	2	2	2		
3	4	4			
4	8				

The net single premium of the annuity is thus $S_1^{\bar{a}} + S_3^{\bar{a}}$, with

$$\begin{aligned}
S_1^{\bar{a}} &= \bar{a}_w + \bar{a}_x + \bar{a}_y + \bar{a}_z , \\
S_3^{\bar{a}} &= \bar{a}_{w:x:y} + \bar{a}_{w:x:z} + \bar{a}_{w:y:z} + \bar{a}_{x:y:z} .
\end{aligned} \qquad (8.5.10)$$

As a second illustration we consider a life insurance for 3 lives (initial ages x, y, z), for which the sum insured is 2 on the first death, 5 on the second death, and 10 on the third death, each payable at the end of the year. The net single premium of this insurance is

$$2\,A_{\overset{3}{x\,:\,y\,:\,z}} + 5\,A_{\overset{2}{x\,:\,y\,:\,z}} + 10\,A_{\overset{}{x\,:\,y\,:\,z}}\,. \tag{8.5.11}$$

Starting with $d_1 = 10$, $d_2 = 5$, $d_3 = 2$ we may complete the difference table:

k	d_k	Δd_k	$\Delta^2 d_k$
1	10	-5	2
2	5	-3	
3	2		

The net single premium of the insurance is thus $10\,S_1^A - 5\,S_2^A + 2\,S_3^A$, with

$$
\begin{aligned}
S_1^A &= A_x + A_y + A_z, \\
S_2^A &= A_{x:y} + A_{x:z} + A_{y:z}, \\
S_3^A &= A_{x:y:z}\,.
\end{aligned}
\tag{8.5.12}
$$

8.6 The Schuette-Nesbitt Formula

Let $B_1, B_2, \cdots, B_m$ denote arbitrary events. Let N denote the number of events that occur; N is a random variable ranging over $\{0, 1, \cdots, m\}$. For arbitrarily chosen coefficients $c_0, c_1, \cdots, c_m$, the formula

$$\sum_{n=0}^{m} c_n \Pr(N = n) = \sum_{k=0}^{m} \Delta^k c_0 S_k \tag{8.6.1}$$

holds, with S_k defined as in (8.4.4), and $S_0 = 1$.

To prove (8.6.1) we use the *shift operator* E defined by

$$E c_k = c_{k+1}\,. \tag{8.6.2}$$

The shift operator and the *difference operator* are connected through the relation $E = 1 + \Delta$. Since $1 - I_{B_j}$ is the indicator function of the complement of B_j, it is easy to see that

$$
\begin{aligned}
\sum_{n=0}^{m} I_{\{N=n\}} E^n &= \prod_{j=1}^{m} (1 - I_{B_j} + I_{B_j} E) \\
&= \prod_{j=1}^{m} (1 + I_{B_j} \Delta) \\
&= \sum_{k=0}^{m} \left(\sum I_{B_{j_1} \cap B_{j_2} \cap \cdots \cap B_{j_k}} \right) \Delta^k\,.
\end{aligned}
\tag{8.6.3}
$$

Taking expectations we obtain the operator identity

$$\sum_{n=0}^{m} \Pr(N = n)E^n = \sum_{k=0}^{m} S_k \Delta^k .$$
(8.6.4)

Applying this operator to the sequence of c_k at $k = 0$, we obtain (8.6.1)

The Schuette-Nesbitt formula (8.6.1) is an elegant and useful generalisation of the much older formulae of Waring, which express $\Pr(N = n)$ and $\Pr(N \geq n)$ in terms of $S_1, S_2, \cdots, S_m$.

Equation (8.5.3) follows from (8.6.1) when B_j is taken to be the event $T_j \geq t$.

Finally we shall present an application which lies outside the field of actuarial mathematics. Letting $c_n = z^n$ in (8.6.1), we obtain an expression for the generating function of N,

$$\mathrm{E}[z^N] = \sum_{k=0}^{m} (z-1)^k S_k .$$
(8.6.5)

Consider as an illustration the following matching problem. Assume that m different letters are inserted into m addressed envelopes at random. Let B_j be the event that letter j is inserted into the correct envelope, and let N be the number of letters with correct address. From

$$\Pr(B_{j_1} \cap B_{j_2} \cap \cdots \cap B_{j_k}) = \frac{1}{m(m-1)\cdots(m-k+1)} ,$$
(8.6.6)

it follows that $S_k = 1/k!$. The generating function of N is thus

$$\mathrm{E}[z^N] = \sum_{k=0}^{m} \frac{(z-1)^k}{k!} .$$
(8.6.7)

For $m \to \infty$ this function converges to e^{z-1}, which is the generating function of the *Poisson distribution* with parameter 1. For large values of m, the distribution of N may thus be approximated by the Poisson distribution with parameter 1.

8.7 Asymmetric Annuities

In general a compound status is less symmetric. For example, the status

$$\overline{w : x} : \overline{y : z}$$
(8.7.1)

is intact, if at least one of (w) and (x) *and* at least one of (y) and (z) survives. The failure time of the status is

$$T = \mathrm{Min}(\mathrm{Max}(T(w), T(x)), \mathrm{Max}(T(y), T(z))) .$$
(8.7.2)

For this status the net single premium of an annuity can be calculated in terms of the net single premiums of joint-life statuses. This follows from the relations

$$_tp_{\overline{u:v}} = {}_tp_u + {}_tp_v - {}_tp_{u:v} , \tag{8.7.3}$$

respectively

$$a_{\overline{u:v}} = a_u + a_v - a_{u:v} , \tag{8.7.4}$$

which are valid for arbitrary statuses u and v. Consider for example an annuity of 1 unit while the status (8.7.1) lasts. By repeated application of (8.7.4) we obtain an expression for the net single premium,

$$
\begin{aligned}
a_{\overline{w:x:y:z}} &= a_{\overline{w:x}:y} + a_{\overline{w:x}:z} - a_{\overline{w:x}:y:z} \\
&= a_{w:y} + a_{x:y} - a_{w:x:y} \\
&\quad + a_{w:z} + a_{x:z} - a_{w:x:z} \\
&\quad - a_{w:y:z} - a_{x:y:z} + a_{w:x:y:z} .
\end{aligned} \tag{8.7.5}
$$

Reversionary annuities are relevant when studying widows' and orphans' insurance. The symbol $\bar{a}_{x/y}$ denotes the net single premium of a continuous payment stream of rate 1, which starts at the death of (x) and terminates at the death of (y). This net single premium can be calculated with the aid of the relation

$$\bar{a}_{x/y} = \bar{a}_y - \bar{a}_{x:y} . \tag{8.7.6}$$

8.8 Asymmetric Insurances

Consider the m lives of Section 8.2 and assume independence of their future lifetimes. A general insurance on the first death provides a benefit of $c_j(t)$ if life j dies first at time t (i.e. the joint-life status fails due to cause j). Such an insurance is mathematically equivalent to the insurance discussed in Section 7.4. In analogy to formula (7.4.4), the net single premium of this first-death insurance is

$$\sum_{j=1}^m \int_0^\infty c_j(t) v^t \, {}_tp_{x_1:x_2:\cdots:x_m} \mu_{x_j+t} dt . \tag{8.8.1}$$

The reversionary annuity considered in the previous section is of this type. Defining

$$c_1(t) = \bar{a}_{y+t} , \quad c_2(t) = 0 , \tag{8.8.2}$$

we obtain

$$\bar{a}_{x/y} = \int_0^\infty \bar{a}_{y+t} v^t \, {}_tp_{x:y} \mu_{x+t} dt . \tag{8.8.3}$$

This expression presupposes independence between $T(x)$ and $T(y)$, in contrast to (8.7.6).

In the special case with $c_k(t) = 1$ and $c_j(t) = 0$ for $j \neq k$, the net single premium is denoted by

$$\bar{A}_{\underset{x_1:\cdots:x_{k-1}:x_k:x_{k+1}:\cdots:x_m}{1}} , \tag{8.8.4}$$

and given by the expression

$$\bar{A}_{\underset{x_1:\cdots:x_{k-1}:x_k:x_{k+1}:\cdots:x_m}{1}} = \int_0^\infty v^t \, {}_tp_{x_1:x_2:\cdots:x_m} \mu_{x_k+t} dt . \tag{8.8.5}$$

Note that the symbols introduced in Chapter 3 to denote the net single premium of a pure endowment, and that of a term insurance, are special cases of (8.8.4); these are obtained by interpreting $\overline{n}$ as a status which fails at time n.

The net single premium (8.8.5) is very easy to calculate if all lives observe the same Gompertz mortality law, see formula (8.3.1). In that case,

$$\mu_{x_k+t} = \frac{c^{x_k}}{c^w} \mu_{x_1+t:x_2+t:\cdots:x_m+t} , \tag{8.8.6}$$

with w defined by (8.3.2); it follows that

$$\bar{A}_{\underset{x_1:\cdots:x_{k-1}:x_k:x_{k+1}:\cdots:x_m}{1}} = \frac{c^{x_k}}{c^w} \bar{A}_{x_1:x_2:\cdots:x_m} = \frac{c^{x_k}}{c^w} \bar{A}_w . \tag{8.8.7}$$

We shall now consider an insurance which pays a benefit of 1 unit at the time of death of (x_k), provided that this is the rth death. Its net single premium is denoted by

$$\bar{A}_{\underset{x_1:\cdots:x_{k-1}:x_k:x_{k+1}:\cdots:x_m}{r}} . \tag{8.8.8}$$

In order that a payment be made at the death of (x_k), exactly $m - r$ of the other $m - 1$ must survive (x_k). Hence we have

$$\bar{A}_{\underset{x_1:\cdots:x_{k-1}:x_k:x_{k+1}:\cdots:x_m}{r}} = \int_0^\infty v^t \, {}_tp_{\underset{x_1:x_2:\cdots:x_{k-1}:x_{k+1}:\cdots:x_m}{[m-r]}} \, {}_tp_{x_k} \mu_{x_k+t} dt . \tag{8.8.9}$$

Substituting as in equation (8.5.3), we obtain a linear combination of net single premiums of the form (8.8.4) is, which makes the calculation easier. Consider for instance

$$\bar{A}_{w:x:y:\overset{2}{z}} = \int_0^\infty v^t \, {}_tp_{\underset{w:x:y}{[2]}} \, {}_tp_z \mu_{z+t} dt . \tag{8.8.10}$$

We now use (8.5.3) with $c_0 = c_1 = c_3 = 0$, $c_2 = 1$ and find that

$${}_tp_{\underset{w:x:y}{[2]}} = S_2^t - 3 S_3^t = {}_tp_{w:x} + {}_tp_{w:y} + {}_tp_{x:y} - 3 \, {}_tp_{w:x:y} . \tag{8.8.11}$$

Substituting the last expression in (8.8.10) yields

$$\bar{A}_{w:x:y:\overset{2}{z}} = \bar{A}_{w:x:\overset{1}{z}} + \bar{A}_{w:y:\overset{1}{z}} + \bar{A}_{x:y:\overset{1}{z}} - 3 \bar{A}_{w:x:y:\overset{1}{z}} . \tag{8.8.12}$$

Chapter 9. The Total Claim Amount in a Portfolio

9.1 Introduction

We consider a certain portfolio of insurance policies and the total amount of claims arising from it during a given period (e.g. a year). We are particularly interested in the probability distribution of the total claim amount, which will allow us to estimate the risk and show whether or not there is a need for reinsurance.

We assume that the portfolio consists of n insurance policies. The claim made in respect of policy h is denoted by S_h. Let us denote the possible values of the random variable S_h by $0, s_{1h}, s_{2h}, \cdots, s_{mh}$, and define

$$\Pr(S_h = 0) = p_h, \quad \Pr(S_h = s_{jh}) = q_{jh} \qquad (9.1.1)$$

for $j = 1, \cdots, m$ and $h = 1, \cdots, n$. With respect to the general insurance type of Chapter 7, q_{jh} may be taken to be the probability of a decrement due to cause j, and s_{jh} may be taken to be the corresponding amount at risk (i.e. the difference between the payment to be made and the available net premium reserve).

The total, or aggregate, amount of claims is

$$S = S_1 + S_2 + \cdots + S_n . \qquad (9.1.2)$$

To enable us to calculate the distribution of S we shall assume that the random variables $S_1, S_2, \cdots, S_n$ are independent.

9.2 The Normal Approximation

The first and second order moments of S may be readily calculated. One has

$$\mathrm{E}[S] = \sum_{h=1}^{n} \mathrm{E}[S_h], \quad \mathrm{Var}[S] = \sum_{h=1}^{n} \mathrm{Var}[S_h], \qquad (9.2.1)$$

with

$$\mathrm{E}[S_h] = \sum_{j=1}^{m} s_{jh} q_{jh}\,, \quad \mathrm{Var}[S_h] = \sum_{j=1}^{m} s_{jh}^2 q_{jh} - \mathrm{E}[S_h]^2\,. \qquad (9.2.2)$$

For a large portfolio (large n) it seems reasonable to approximate the probability distribution of S by a normal distribution with parameters $\mu = \mathrm{E}[S]$ and $\sigma^2 = \mathrm{Var}[S]$. However, the quality of this approximation depends not only on the size of the portfolio, but also on its homogeneity. Moreover, this approximation is not uniformly good: in general the results are good around the mean $\mathrm{E}[S]$ and less satisfactory in the "tails" of the distribution.

These weaknesses of the approximation by the normal distribution may be partially relieved by sophisticated procedures, such as the *Esscher Method* or the *Normal Power Approximation*. However, these methods have lost some of their interest: if a high-powered computer is available, the distribution of S can be calculated more or less "exactly".

9.3 Exact Calculation of the Total Claim Amount Distribution

The probability distribution of S is obtained by the convolution of the probability distributions of $S_1, \cdots, S_n$. The distributions of S_1+S_2, $S_1+S_2+S_3$, $S_1+S_2+S_3+S_4, \cdots$, are found successively. If the distribution of $S_1 + \cdots + S_{h-1}$ is known, the distribution of $S_1 + \cdots + S_h$ may be calculated by the formula

$$\begin{aligned}
\mathrm{Pr}(S_1 + \cdots + S_h = x) \;=\; & \sum_{j=1}^{m} \mathrm{Pr}(S_1 + \cdots + S_{h-1} = x - s_{jh}) q_{jh} \\
& + \mathrm{Pr}(S_1 + \cdots + S_{h-1} = x) p_h\,. \qquad (9.3.1)
\end{aligned}$$

With this procedure it is desirable that the s_{jh} are multiples of some basic monetary unit. Of course, in general this will not be the case unless the basic monetary unit is chosen very small. The original distribution of S_h is then appropriately modified. Two methods are popular in this respect.

Method 1 (Rounding)

The method starts by replacing s_{jh} by a rounded value s_{jh}^*, which is a multiple of the chosen monetary unit. In order to keep the expected total claim amount the same the probabilities are adjusted accordingly by the substitutions:

$$s_{jh} \to s_{jh}^*\,,\; q_{jh} \to q_{jh}^* = q_{jh} s_{jh}/s_{jh}^*,\; p_h \to p_h^* = 1 - (q_{1h}^* + \cdots + q_{mh}^*)\,. \quad (9.3.2)$$

Method 2 (Dispersion)

Let s_{jh}^- denote the largest multiple (of the desired monetary unit) not exceeding s_{jh}, and let s_{jh}^+ denote the least multiple which exceeds s_{jh}. The original distribution of S_h has a point mass of q_{jh} at s_{jh}. The dispersion method consists of re-allocating this point mass to s_{jh}^- and s_{jh}^+ in such a way that the expectation is unchanged. The new point masses q_{jh}^- and q_{jh}^+ must therefore satisfy the equations

$$q_{jh}^- + q_{jh}^+ = q_{jh} , \quad s_{jh}^- q_{jh}^- + s_{jh}^+ q_{jh}^+ = s_{jh} q_{jh} , \qquad (9.3.3)$$

that is

$$q_{jh}^- = \frac{s_{jh}^+ - s_{jh}}{s_{jh}^+ - s_{jh}^-} q_{jh} , \quad q_{jh}^+ = \frac{s_{jh} - s_{jh}^-}{s_{jh}^+ - s_{jh}^-} q_{jh} . \qquad (9.3.4)$$

Consider as an illustration a portfolio of three policies with, for example:

$$\begin{aligned}
&\Pr(S_1 = 0) = 0.8, \quad \Pr(S_1 = 0.5) = 0.1, \quad \Pr(S_1 = 2.5) = 0.1, \\
&\Pr(S_2 = 0) = 0.7, \quad \Pr(S_2 = 1.25) = 0.2, \quad \Pr(S_2 = 2.5) = 0.1, \qquad (9.3.5) \\
&\Pr(S_3 = 0) = 0.6, \quad \Pr(S_3 = 1.5) = 0.2, \quad \Pr(S_3 = 2.75) = 0.2 .
\end{aligned}$$

The convolution of the three distributions ranges over the values 0, 0.5, 1.25, 1.5, 1.75, 2, 2.5, 2.75, $\cdots$, 6.5, 7.75, and it may in principle be calculated. Calculating the convolution of the modified distributions is much easier, however. We shall use Method 2 to approximate the distribution of S_h by a distribution on the integers. The modifications prescribed by Method 2 are set out in the table below:

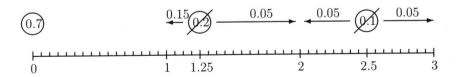

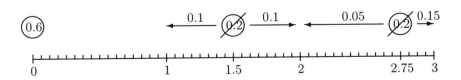

Hence, the modified distributions are as follows:

	$x = 0$	$x = 1$	$x = 2$	$x = 3$
$\Pr(S_1 = x)$	0.85	0.05	0.05	0.05
$\Pr(S_2 = x)$	0.70	0.15	0.10	0.05
$\Pr(S_3 = x)$	0.60	0.10	0.15	0.15

Application of (9.3.1) with $h = 2$ and $h = 3$ yields the distribution of $S = S_1 + S_2 + S_3$ in two steps:

x	$\Pr(S_1 + S_2 = x)$	$\Pr(S_1 + S_2 + S_3 = x)$	$\Pr(S \le x)$
0	0.5950	0.357000	0.357000
1	0.1625	0.157000	0.514000
2	0.1275	0.182000	0.696000
3	0.0900	0.180375	0.876375
4	0.0150	0.061500	0.937875
5	0.0075	0.038625	0.976500
6	0.0025	0.018000	0.994500
7		0.003625	0.998125
8		0.001500	0.999625
9		0.000375	1.000000

In a realistic portfolio (say a pension fund with 1000 members) the original distribution of the S_h will always have to be modified beforehand. In order to keep the notation simple we shall assume that the distribution in (9.1.1) has already been modified, and that the s_{jh} are integers (one may always achieve this by proper choice of the monetary unit). Thus we may simply assume that $s_{jh} = j$, keeping in mind the possibility that some q_{jh} vanish.

9.4 The Compound Poisson Approximation

Assume that the distribution of S_h is given by

$$\Pr(S_h = 0) = p_h, \quad \Pr(S_h = j) = q_{jh} \quad (j = 1, 2, \cdots, m).$$ (9.4.1)

The generating function of this distribution is

$$p_h + \sum_{j=1}^{m} q_{jh} z^j = 1 + \sum_{j=1}^{m} q_{jh}(z^j - 1).$$ (9.4.2)

The distribution of S_h may now be approximated by the corresponding compound Poisson distribution whose generating function is

$$g_h(z) = \exp\left(\sum_{j=1}^m q_{jh}(z^j - 1)\right).$$ (9.4.3)

By comparing (9.4.2) and (9.4.3) one will see that the approximation is best for small values of the q_{jh}.

If we now use the compound Poisson approximation for all terms in (9.1.2), the resulting approximation of S will have as generating function

$$g(z) = \prod_{h=1}^n g_h(z) = \exp\left(\sum_{j=1}^m q_j(z^j - 1)\right),$$ (9.4.4)

with the notation

$$q_j = \sum_{h=1}^n q_{jh}.$$ (9.4.5)

But this means that the distribution of S can also be approximated by a compound Poisson distribution. In the corresponding model the total claim amount is

$$S = X_1 + X_2 + \cdots + X_N;$$ (9.4.6)

here N denotes the random number of claims, and X_i denotes the amount of the ith claim. Furthermore, the random variables $N, X_1, X_2, \cdots$ are independent, N has a Poisson distribution with parameter

$$q = q_1 + q_2 + \cdots + q_m,$$ (9.4.7)

and the probability that the amount of an individual claim is j is

$$p(j) = q_j/q \quad (j = 1, 2, \cdots, m).$$ (9.4.8)

The probability distribution of S is then given by the formula

$$\Pr(S = x) = \sum_{k=0}^\infty p^{*k}(x)e^{-q}q^k/k!.$$ (9.4.9)

In the numerical example in the previous section we had $q_1 = 0.3$, $q_2 = 0.3$, $q_3 = 0.25$. Thus $q = 0.85$, and each of the random variables X_i may take the values 1,2 or 3, with probabilities $p(1) = 30/85$, $p(2) = 30/85$, $p(3) = 25/85$.

The model (9.4.6), called the *collective risk model*, is particularly appropriate if the portfolio is subject to changes during the year. Even in such a dynamic portfolio it will be possible to estimate the expected number of claims (q) and the individual claim amount distribution.

Note that (9.4.6) can be written as

$$S = N_1 + 2N_2 + 3N_3 + \cdots + mN_m,$$ (9.4.10)

if we let N_j denote the number of claims for amount j. It can be proved that the random variables $N_1, N_2, \cdots, N_m$ are independent, and that N_j has a Poisson distribution with parameter q_j (so that q_j is the frequency of claims for amount j).

The distribution of S can in principle be calculated by either (9.4.9) or (9.4.10). A third method, the *recursive method*, will be presented in the next section.

9.5 Recursive Calculation of the Compound Poisson Distribution

Let us denote the probabilities $\Pr(S = x)$ by $f(x)$ and the cumulative distribution function by $F(x) = \Pr(S \leq x)$. Thus, for example,

$$f(0) = \Pr(S = 0) = \Pr(N = 0) = e^{-q}. \qquad (9.5.1)$$

Panjer directed the attention of actuaries to the useful recursive formula

$$f(x) = \frac{1}{x} \sum_{j=1}^{m} j q_j f(x - j), \quad x = 1, 2, 3, \cdots, \qquad (9.5.2)$$

which enables us to calculate the values $f(1), f(2), f(3), \cdots$ successively.

In the numerical example considered above the calculations are as follows:

$$
\begin{aligned}
f(0) &= e^{-0.85}, \\
f(1) &= 0.3\, f(0), \\
f(2) &= \frac{1}{2}(0.3\, f(1) + 0.6\, f(0)), \\
f(3) &= \frac{1}{3}(0.3\, f(2) + 0.6\, f(1) + 0.75\, f(0)), \\
f(4) &= \frac{1}{4}(0.3\, f(3) + 0.6\, f(2) + 0.75\, f(1)), \\
&\cdots
\end{aligned}
\qquad (9.5.3)
$$

The numerical results have been compiled in the following table; the partial sums $F(x)$ could, of course, also have been calculated recursively.

x	$f(x)$	$F(x)$	x	$f(x)$	$F(x)$
0	0.427415	0.427415	10	0.001302	0.998886
1	0.128224	0.555639	11	0.000645	0.999531
2	0.147458	0.703098	12	0.000277	0.999808
3	0.147244	0.850342	13	0.000111	0.999920
4	0.057204	0.907546	14	0.000049	0.999969
5	0.043220	0.950766	15	0.000019	0.999988
6	0.026287	0.977053	16	0.000007	0.999995
7	0.010960	0.988014	17	0.000003	0.999998
8	0.006434	0.994448	18	0.000001	0.999999
9	0.003136	0.997584			

The generating function of S can be used to prove the recursive formula (9.5.2). On the one hand, it is defined by

$$g(z) = \sum_{x=0}^{\infty} f(x)z^x \,, \tag{9.5.4}$$

while, on the other hand, (9.4.4) implies that

$$\ln g(z) = \sum_{j=1}^{m} q_j(z^j - 1) \,. \tag{9.5.5}$$

From the identity

$$\frac{d}{dz}g(z) = g(z)\frac{d}{dz}\ln g(z) \tag{9.5.6}$$

we obtain

$$\sum_{x=1}^{\infty} xf(x)z^{x-1} = \left(\sum_{y=0}^{\infty} f(y)z^y\right)\left(\sum_{j=1}^{m} jq_j z^{j-1}\right) \,. \tag{9.5.7}$$

Comparing the coefficients of z^{x-1}, we find

$$xf(x) = \sum_{j=1}^{m} f(x-j)jq_j \,, \tag{9.5.8}$$

which establishes (9.5.2)

Until now we have tacitly assumed that only positive claims could occur, that is that all terms in (9.4.6) are positive. If negative claims can occur, the total amount of claims can be decomposed into S^+, the sum of positive claims, and S^-, the sum of the absolute values of the negative claims:

$$S = S^+ - S^- \tag{9.5.9}$$

It can be shown that both S^+ and S^- have compound Poisson distributions and are independent. We can now compute the distributions of S^+ and S^- separately, e.g. from (9.5.2), and finally obtain the distribution of S by convolution.

9.6 Reinsurance

If inspection of the distribution of S shows that the risk is too high the acquisition of proper reinsurance is indicated. Different forms of reinsurance are available, two of which will be discussed in this and the next section.

Quite generally a reinsurance contract guarantees the insurer the reimbursement of an amount R (a function of the individual claims and thus a random variable) in return for a reinsurance premium Π. The insurer's *retention* is

$$\hat{S} = S + \Pi - R. \tag{9.6.1}$$

With proper reinsurance the distribution of $\hat{S}$ will be more favourable than the distribution of S. Let us define $\hat{f}(x) = \Pr(\hat{S} = x)$ and $\hat{F}(x) = \Pr(\hat{S} \leq x)$.

An *Excess of Loss* reinsurance with priority α reimburses the excess $X_i - \alpha$ for all individual claims which exceed α.

Let us assume in our numerical example that excess of loss reinsurance with $\alpha = 1$ can be purchased for a premium of $\Pi = 1.2$. The original claims which can assume the values 1,2,3, are all reduced to 1 by the reinsurance arrangement. Thus the insurer's retention is

$$\hat{S} = 1.2 + N ; \tag{9.6.2}$$

here N denotes the number of claims and has a Poisson distribution with parameter 0.85. The distribution of $\hat{S}$ is tabulated below:

x	$\hat{f}(x)$	$\hat{F}(x)$
1.2	0.427415	0.427415
2.2	0.363303	0.790718
3.2	0.154404	0.945121
4.2	0.043748	0.988869
5.2	0.009296	0.998165
6.2	0.001580	0.999746
7.2	0.000224	0.999970
8.2	0.000027	0.999997
9.2	0.000003	1.000000

Since the reinsurance premium contains a loading, $\Pi > E[R]$, it is clear from (9.6.1) that $E[\hat{S}] > E[S]$; in our example we have $E[\hat{S}] = 2.05$, while $E[S] = 1.65$. The purpose of reinsurance is to reduce the probabilities of large total claims; indeed in our example we have $\hat{F}(6.2) = 0.999746$, which exceeds the corresponding probability without reinsurance by far $(F(6) = 0.977053)$. In the next section we shall present a reinsurance form which is extremely effective in this respect.

9.7 Stop-Loss Reinsurance

Under a stop-loss reinsurance contract with deductible β, the excess $R = (S - \beta)^+$ of the total claims over the specified deductible is reimbursed. In this case

$$\hat{S} = S + \Pi - (S - \beta)^+ = \begin{cases} S + \Pi & \text{if } S < \beta, \\ \beta + \Pi & \text{if } S \geq \beta. \end{cases} \qquad (9.7.1)$$

Let us now assume that a stop-loss cover for the deductible $\beta = 3$ has been bought at a premium of $\Pi = 1.1$. The insurer's portion of the total claim amount will be limited to 3. The distribution of $\hat{S}$ can be derived from the distribution of S:

x	$\hat{f}(x)$	$\hat{F}(x)$
1.1	0.427415	0.427415
2.1	0.128224	0.555639
3.1	0.147458	0.703098
4.1	0.296903	1.000000

The expected value of $\hat{S}$ is quite large, $E[\hat{S}] = 2.41$, but the "risk" has been reduced to a minimum.

We shall finally consider calculation of the net stop-loss premium, which we denote by $\rho(\beta)$:

$$\rho(\beta) = E[(S - \beta)^+] = \int_\beta^\infty (x - \beta)dF(x). \qquad (9.7.2)$$

By partial integration we obtain

$$\rho(\beta) = \int_\beta^\infty [1 - F(x)]dx. \qquad (9.7.3)$$

Hence, for integer values of β, we may write

$$\rho(\beta) = \sum_{x=\beta}^\infty [1 - F(x)], \qquad (9.7.4)$$

or, written recursively,

$$\rho(\beta + 1) = \rho(\beta) - [1 - F(\beta)]. \qquad (9.7.5)$$

Thus the values $\rho(1), \rho(2), \rho(3), \cdots$ can be computed successively, starting with $\rho(0) = E[S]$. Of course, these computations can be combined with the recursive calculation of $F(x)$ (see Section 9.5).

In our example the stop-loss premiums assume the following values.

β	$\rho(\beta)$
0	1.650000
1	1.077415
2	0.633054
3	0.336152
4	0.186494
5	0.094040
6	0.044807
7	0.021860
8	0.009874
9	0.004322
10	0.001906
...	...

Of course, the actual stop-loss premium Π will exceed the net premium $\rho(\beta)$ significantly. Our example, with $\Pi = 1.1$ and $\rho(3) = 0.336152$ corresponds to a 227% loading. Loadings of this order of magnitude are not uncommon.

The net premium is still of interest, since it allows one to calculate the expected value of the retention, which is

$$E[\hat{S}] = E[S] + \Pi - \rho(\beta) \, . \tag{9.7.6}$$

In our example we have again $E[\hat{S}] = 1.65 + 1.1 - 0.34 = 2.41$.

Chapter 10. Expense Loadings

10.1 Introduction

The operations of an insurance contract will involve certain expenses, whether undertaken by pension funds or by insurance companies. In the case of a pension fund these expenses are most often lumped together and considered separately from the strictly technical insurance analysis. In the case of insurance companies, on the other hand, the cost element is built into the model, as explicitly and equitably as possible. As we shall see, however, the resulting premiums and reserves are very closely related to the net premiums and reserves we have been discussing so far, and which will therefore continue to hold our primary interest.

Expenses can be classified into three main groups:

a. Acquisition Expenses

These comprise all expenses connected with a new policy issue: agents' commission and travel expenses, medical examination, policy writing, advertising. These expenses are charged against the policy as a single amount, which is proportional to the sum insured. The corresponding rate will be denoted by α.

b. Collection Expenses

These expenses are charged at the beginning of every year in which a premium is to be collected. We assume that these expenses are proportional to the expense-loaded premium (see 10.2), at a rate which we will denote by β.

c. Administration Expenses

All other expenses are included in this item, such as wages, rents, data processing costs, investment costs, taxes, license fees etc. These costs are charged against the policy during its entire contract period, at the beginning of every policy year, usually as a proportion of the sum insured, respectively the annuity level, and the corresponding rate is denoted by γ.

This traditional allocation of expenses is somewhat arbitrary. Some expense items will obviously be fixed costs, independent of the sum insured. Nevertheless, the assumption of proportionality is retained for the sake of simplicity. The factors α, β and γ will, however, depend on the type of insurance involved. Expenses in respect of an individual insurance are relatively higher than expenses in respect of a group insurance; for the latter the acquisition expense is often even waived entirely (i.e. $\alpha = 0$).

10.2 The Expense-Loaded Premium

The *expense-loaded premium* (or *adequate premium*) , which we will denote by P^a, is the amount of annual premium of which the expected present value is just sufficient to finance the insured benefits, as well as the incurred costs in respect of the insurance policy. Hence we may write

$$P^a = P + P^\alpha + P^\beta + P^\gamma\,; \tag{10.2.1}$$

here P denotes the net annual premium, while P^α, P^β and P^γ denote the three components of the expense loading.

We consider as a first example an endowment (sum insured: 1, duration: n, age at issue: x). The expense-loaded annual premium must satisfy the condition

$$P^a_{x:\overline{n}|}\,\ddot{a}_{x:\overline{n}|} = A_{x:\overline{n}|} + \alpha + \beta\,P^a_{x:\overline{n}|}\,\ddot{a}_{x:\overline{n}|} + \gamma\,\ddot{a}_{x:\overline{n}|}, \tag{10.2.2}$$

so that

$$P^a_{x:\overline{n}|} = \frac{A_{x:\overline{n}|} + \alpha + \gamma\ddot{a}_{x:\overline{n}|}}{(1-\beta)\ddot{a}_{x:\overline{n}|}}\,. \tag{10.2.3}$$

The expense-loaded annual premium will be expressed in terms of the net annual premium if we replace α by $\alpha(A_{x:\overline{n}|} + d\ddot{a}_{x:\overline{n}|})$ in the above formula:

$$P^a_{x:\overline{n}|} = \frac{1+\alpha}{1-\beta}P_{x:\overline{n}|} + \frac{\alpha d + \gamma}{1-\beta}\,. \tag{10.2.4}$$

If we now divide (10.2.2) by $\ddot{a}_{x:\overline{n}|}$, we obtain (10.2.1) in the specific form:

$$P^a_{x:\overline{n}|} = P_{x:\overline{n}|} + \frac{\alpha}{\ddot{a}_{x:\overline{n}|}} + \beta P^a_{x:\overline{n}|} + \gamma\,. \tag{10.2.5}$$

As a second example we consider the same endowment, but with a shorter premium paying period $m < n$. The expense-loaded annual premium is obtained from the condition

$$P^a\ddot{a}_{x:\overline{m}|} = A_{x:\overline{n}|} + \alpha + \beta P^a\ddot{a}_{x:\overline{m}|} + \gamma\ddot{a}_{x:\overline{n}|}. \tag{10.2.6}$$

Its components are

$$P^a = P + \frac{\alpha}{\ddot{a}_{x:\overline{m}|}} + \beta P^a + \gamma \frac{\ddot{a}_{x:\overline{n}|}}{\ddot{a}_{x:\overline{m}|}}, \qquad (10.2.7)$$

with, of course, $P = A_{x:\overline{n}|}/\ddot{a}_{x:\overline{m}|}$.

For deferred annuities financed by annual premiums it is customary to charge acquisition expenses as a fraction of the expense-loaded annual premium, in the same way as collection expenses. Here it is also possible to use two administration expense rates, a rate γ_1 for the premium paying period, and another rate γ_2 for the annuity's duration.

For simplicity, the reader may identify the expense-loaded premium with the gross premium; the necessary safety loading is then taken to be implicit in the "net" premium, through conservative assumptions about interest and mortality rates. In practice, the gross premium may also differ from the expense-loaded premium in that either surcharges for small policies or discounts for large policies are used.

In some countries the premium quoted by the insurance company consists of the net premium and administration expenses, but not acquisition and collection expenses. This premium (German: Inventarprämie),

$$P^{inv} = P + P^\gamma, \qquad (10.2.8)$$

covers the actual costs of insured benefits and internal administration expenses.

10.3 Expense-Loaded Premium Reserves

The expense-loaded premium reserve (or adequate reserve) at the end of year k is denoted by ${}_kV^a$. It is defined as the difference between the expected present value of future benefits plus expenses, and that of future expense-loaded premiums. The expense-loaded premium reserve can be separated into components similar to those of the expense-loaded premium:

$${}_kV^a = {}_kV + {}_kV^\alpha + {}_kV^\gamma. \qquad (10.3.1)$$

Here ${}_kV$ denotes the net premium reserve, ${}_kV^\alpha$ is the negative of the expected present value of future P^α, and the *reserve for administration expenses* is the difference in expected present value between future administration expenses and future P^γ.

For an endowment we have

$$\begin{aligned}
{}_kV^\alpha_{x:\overline{n}|} &= -P^\alpha \ddot{a}_{x+k:\overline{n-k}|} \\
&= -\alpha \frac{\ddot{a}_{x+k:\overline{n-k}|}}{\ddot{a}_{x:\overline{n}|}} \\
&= -\alpha(1 - {}_kV_{x:\overline{n}|}) \qquad (10.3.2)
\end{aligned}$$

and $_k V^\gamma = 0$ for $k = 1, 2, \cdots, n$. Thus

$$_k V^a_{x:\overline{n}|} = (1 + \alpha) \, _k V_{x:\overline{n}|} - \alpha . \tag{10.3.3}$$

If the premium paying period is reduced to m years, then

$$_k V^\alpha = -P^\alpha \, \ddot{a}_{x+k:\overline{m-k}|} = -\alpha(1 - \, _k V_{x:\overline{m}|}) \tag{10.3.4}$$

for $k = 1, 2, \cdots, m-1$, and $_k V^\alpha = 0$ for $k \geq m$. The reserve for administration expenses is then

$$
\begin{aligned}
_k V^\gamma &= \gamma \, \ddot{a}_{x+k:\overline{n-k}|} - P^\gamma \, \ddot{a}_{x+k:\overline{m-k}|} \\
&= \gamma \, \ddot{a}_{x:\overline{n}|} \left(\frac{\ddot{a}_{x+k:\overline{n-k}|}}{\ddot{a}_{x:\overline{n}|}} - \frac{\ddot{a}_{x+k:\overline{m-k}|}}{\ddot{a}_{x:\overline{m}|}} \right) \\
&= \gamma \, \ddot{a}_{x:\overline{n}|} (\, _k V_{x:\overline{m}|} - \, _k V_{x:\overline{n}|})
\end{aligned}
\tag{10.3.5}
$$

for $k = 1, 2, \cdots, m-1$, and

$$_k V^\gamma = \gamma \, \ddot{a}_{x+k:\overline{n-k}|} \tag{10.3.6}$$

for $k \geq m$.

The idea to include the negative acquisition cost reserve $_k V^\alpha$ in the premium reserve is due to *Zillmer*. In the first few years, the expense-loaded premium reserve may be negative if α is large. Hence the need for upper bounds on α arose. One suggestion was to choose the value of α at most equal to the one for which the expense-loaded premium reserve is zero at the end of the first year. Consider an endowment as an illustration. The condition $_1 V^a_{x:\overline{n}|} \geq 0$ together with (10.3.3) implies that the acquisition expense rate cannot exceed

$$\alpha = \, _1 V_{x:\overline{n}|}/(1 - \, _1 V_{x:\overline{n}|}) . \tag{10.3.7}$$

With the substitutions

$$_1 V_{x:\overline{n}|} = (P_{x+1:\overline{n-1}|} - P_{x:\overline{n}|}) \, \ddot{a}_{x+1:\overline{n-1}|}, \tag{10.3.8}$$

and

$$1 - \, _1 V_{x:\overline{n}|} = \ddot{a}_{x+1:\overline{n-1}|}/ \, \ddot{a}_{x:\overline{n}|}, \tag{10.3.9}$$

the upper bound becomes

$$\alpha = (P_{x+1:\overline{n-1}|} - P_{x:\overline{n}|}) \, \ddot{a}_{x:\overline{n}|}. \tag{10.3.10}$$

Thus it is evident that

$$P + P^\alpha = P_{x:\overline{n}|} + \frac{\alpha}{\ddot{a}_{x:\overline{n}|}} = P_{x+1:\overline{n-1}|}. \tag{10.3.11}$$

This result should not come as a surprise: Since $_1V + {}_1V^\alpha = 0$, the premiums of $P + P^\alpha$ paid from age $x + 1$ and onward must be sufficient to cover the future benefits. It is also clear that then

$$_kV_{x:\overline{n}|} + {}_kV^\alpha_{x:\overline{n}|} = {}_{k-1}V_{x+1:\overline{n-1}|} \tag{10.3.12}$$

holds for $k = 2, 3, \cdots, n$.

In practical insurance, the maximum value of α is usually given as a fixed percentage (say $\alpha = 3\frac{1}{2}\%$).

In some countries the expense-loaded premium reserve does not include an acquisition cost reserve. The modified expense-loaded reserve (German: Inventardeckungskapital) then becomes

$$_kV^{inv} = {}_kV + {}_kV^\gamma . \tag{10.3.13}$$

Chapter 11. Estimating Probabilities of Death

11.1 Problem Description

The one-year probability of death q_x has to be estimated from statistical data; these data will be generated by a certain group of lives (e.g. policyholders), which has been under observation for a certain period (one or more calendar years), the *observation period*. The estimated value of q_x will be denoted by $\hat{q}_x$.

If all observations are complete, meaning that each life has been observed from age x until age $x + 1$ or prior death, the statistical analysis is quite simple. Unfortunately, this is in practice not the case, as will be illustrated by the so-called *Lexis diagram*:

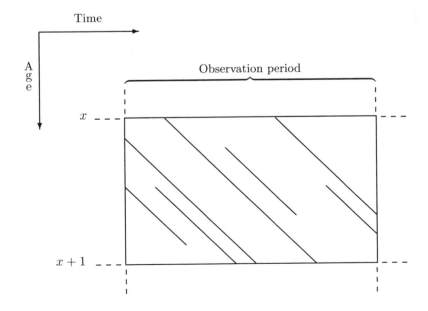

In this diagram each life under observation corresponds to a diagonal line segment showing the time interval during which the life has been observed. The horizontal borders of the rectangle are made up by the age group under consideration, and the vertical borders represent beginning and end of the observation period. Lives aged x before the observation period begins are incompletely observed (some may have died without this being observed); similarly, lives aged $x+1$ after the observation period ends will be incompletely observed. Another source of incomplete observations is lives which enter the group between the ages of x and $x + 1$, when they buy an insurance policy; as well as lives leaving the group between the ages of x and $x + 1$ for reasons other than death, such as policy termination.

Let n lives contribute to the observations in the rectangle. Assume that life no. i is observed between the ages of $x + t_i$ and $x + s_i$ $(0 \le t_i < s_i \le 1)$. The sum

$$E_x = (s_1 - t_1) + (s_2 - t_2) + \cdots + (s_n - t_n) \tag{11.1.1}$$

is called the the *exposure*. The total length of all line segments in the Lexis diagram is $\sqrt{2}\, E_x$.

Let D_x denote the number of deaths observed in the rectangle (unlike E_x, D_x is of course an integer). Denote by I the set of observations i which were terminated by death, and define, for $i \in I$,

$$s_i^{(m)} = [ms_i + 1]/m\,, \tag{11.1.2}$$

i.e. $s_i^{(m)}$ is obtained by rounding s_i to the next mth part of the year.

11.2 The Classical Method

The idea behind the classical method is to equate the expected number of deaths to the observed number of deaths in order to derive an estimator $\hat{q}_x$.

The expected number of deaths is in some sense

$$\sum_{i=1}^{n} {}_{1-t_i}q_{x+t_i} - \sum_{i \notin I} {}_{1-s_i}q_{x+s_i}\,. \tag{11.2.1}$$

This expression is simplified by *Assumption c* of Section 2.6, which states that ${}_{1-u}q_{x+u} = (1 - u)q_x$. The expected number of deaths then becomes

$$\sum_{i=1}^{n}(1 - t_i)q_x - \sum_{i \notin I}(1 - s_i)q_x = E_x q_x + \sum_{i \in I}(1 - s_i)q_x\,. \tag{11.2.2}$$

Equating this expression to the observed number of deaths, we obtain the classical estimator

$$\hat{q}_x = \frac{D_x}{E_x + \sum_{i \in I}(1 - s_i)}\,. \tag{11.2.3}$$

This estimator works well if the volume of data is large. The denominator is sometimes approximated. For instance, under the assumption that deaths, on the average, occur at age $x + \frac{1}{2}$, the estimator is simply

$$\hat{q}_x = \frac{D_x}{E_x + \frac{1}{2} D_x} . \tag{11.2.4}$$

The estimator (11.2.3) does not work satisfactorily with sparse data. One problem is that the numerator may exceed the denominator, giving an obviously useless estimate of q_x; another is that the estimator is not amenable to confidence estimation or hypothesis testing, since its statistical properties are hard to evaluate. Alternative suggestions will be presented below.

11.3 Alternative Solution

Let m be a positive integer, and define $h = 1/m$. We shall estimate $_h q_x$ using the method of the previous section. To this end we assume that $_{h-u} q_{x+u}$ is a linear function of u, i.e.

$$_{h-u} q_{x+u} = (1 - mu) \, _h q_x \quad \text{for } 0 \le u \le h . \tag{11.3.1}$$

In order to make use of all data we also assume that the force of mortality between the ages of x and $x + 1$ is a periodic function with period h. This assumption implies, for $j = 1, 2, \cdots, m - 1$, that

$$_{h-u} q_{x+jh+u} = \, _{h-u} q_{x+u} \quad \text{for } 0 \le u \le h . \tag{11.3.2}$$

Making use of the two assumptions, one may now argue that the expected number of deaths is

$$m E_x \, _h q_x + m \sum_{i \in I} (s_i^{(m)} - s_i) \, _h q_x . \tag{11.3.3}$$

Equating this to the observed number of deaths, one obtains the estimator

$$_h \hat{q}_x = \frac{h D_x}{E_x + \sum_{i \in I} (s_i^{(m)} - s_i)} . \tag{11.3.4}$$

Assumption (11.3.2) implies that $p_x = (_h p_x)^m$. An estimator of q_x is thus obtained from (11.3.4) by

$$\hat{q}_x = 1 - (1 - \, _h \hat{q}_x)^m . \tag{11.3.5}$$

This alternative procedure does not become interesting until we let $m \to \infty$. In the limiting case the assumptions (11.3.1) and (11.3.2) coincide with *Assumption b* in Section 2.6, stating $\mu_{x+u} = \mu_{x+\frac{1}{2}}$ for $0 < u < 1$, and the expected number of deaths (11.3.3) becomes $E_x \mu_{x+\frac{1}{2}}$. This leads us to estimate the constant value of the force of mortality by the ratio

$$\hat{\mu}_{x+\frac{1}{2}} = \frac{D_x}{E_x}. \tag{11.3.6}$$

The probability q_x is then estimated by

$$\hat{q}_x = 1 - \exp(-\hat{\mu}_{x+\frac{1}{2}}) = 1 - \exp(-D_x/E_x). \tag{11.3.7}$$

11.4 The Maximum Likelihood Method

The moment method of the previous sections may be criticised on the grounds that equating "expected" number of deaths in the expressions (11.2.1), (11.2.2) and (11.3.3) to the observed number of deaths, is a heuristic approach. However, the estimators (11.3.6) and (11.3.7) can also be derived by a different method.

We assume that the n lives are independent. The likelihood function of the observations is then

$$\prod_{i \in I} \mu_{x+s_i} \, {}_{s_i-t_i}p_{x+t_i} \cdot \prod_{i \notin I} {}_{s_i-t_i}p_{x+t_i}. \tag{11.4.1}$$

The assumption of a piecewise constant force of mortality simplifies this to

$$(\mu_{x+\frac{1}{2}})^{D_x} \exp(-\mu_{x+\frac{1}{2}} E_x). \tag{11.4.2}$$

This expression is maximised by $\hat{\mu}_{x+\frac{1}{2}} = D_x/E_x$, so that (11.3.6) is also the maximum likelihood estimator. The invariance principle then implies that $\hat{q}_x$ defined by (11.3.7) will also be the maximum likelihood estimator of q_x.

11.5 Statistical Inference

Actually, both D_x and E_x are random variables. However, it is convenient to treat E_x as a non-random quantity. Let us therefore assume that the random variable D_x has a Poisson distribution with mean

$$\lambda = \mu_{x+\frac{1}{2}} E_x, \tag{11.5.1}$$

with unknown parameter $\mu_{x+\frac{1}{2}}$. The probability of D_x deaths, apart from a factor which is independent of $\mu_{x+\frac{1}{2}}$, then is identical with the likelihood (11.4.2). The point estimators (11.3.6) and (11.3.7) therefore retain their validity.

It is also possible to treat D_x as a non-random quantity, assuming that E_x follows a *gamma distribution* with parameters $\alpha = D_x$ and $\beta = \mu_{x+\frac{1}{2}}$. This approach is also compatible with the likelihood (11.4.2); we shall not pursue this here.

The following table displays confidence limits for the parameter of a Poisson distribution, for an observed value of n. The lower limit λ^l is defined in such a way that the probability of an observation of n or greater, calculated for the value λ^l, is equal to w; similarly, the probability of observing n or less for λ^u is equal to w.

The confidence interval for λ may be read off directly in the table from the number of observed deaths D_x. Dividing the confidence limits by E_x, the confidence interval for $\mu_{x+\frac{1}{2}}$ is obtained. Finally the limits may be transformed to give a confidence interval for q_x. As an illustration, assume that $D_x = 19$ and $E_x = 2000$. The 90% confidence intervals are then $12.44 < \lambda < 27.88$, $0.00622 < \mu_{x+\frac{1}{2}} < 0.01394$, $0.00620 < q_x < 0.01384$.

The estimated probabilities $\hat{q}_x$ (called "crude" rates in practice) may fluctuate wildly from one age interval to the next. In such a situation one may use one of the more or less sophisticated methods of graduation theory in order to obtain a smooth function. We shall not discuss these methods in this book.

It is also possible to use an existing life table as a standard and to postulate that the forces of mortality in the observed group are a constant (age independent) multiple of the forces of mortality in the standard life table. Denoting the forces of mortality in the standard table by $\mu^t_{x+\frac{1}{2}}$, we thus assume that

$$\mu_{x+\frac{1}{2}} = f \mu^t_{x+\frac{1}{2}}, \tag{11.5.2}$$

the objective now being to estimate the factor f. Under the assumption that the number of deaths occuring in different age groups are independent random variables, we see that the total number of deaths,

$$D = \sum_x D_x, \tag{11.5.3}$$

follows a Poisson distribution with mean

$$\lambda = \sum_x \mu_{x+\frac{1}{2}} E_x = f \sum_x \mu^t_{x+\frac{1}{2}} E_x. \tag{11.5.4}$$

The estimator for λ is then $\hat{\lambda} = D$, and we find

$$\hat{f} = \frac{D}{\sum_x \mu^t_{x+\frac{1}{2}} E_x}; \tag{11.5.5}$$

this expression is referred to as the *mortality ratio*. A confidence interval for λ may easily be transformed into a confidence interval for f.

For instance, assume that a total of

$$D = D_{40} + D_{41} + \cdots + D_{49} = 932 \tag{11.5.6}$$

deaths have been observed in the age group between 40 and 50, while the expected number of deaths according to a standard table is

$$\sum_{x=40}^{49} \mu_{x+\frac{1}{2}}^{t} E_x = 1145.7 \,. \tag{11.5.7}$$

Then one obtains $\hat{f} = 932/1145.7 = 0.813 = 81.3\%$. In order to construct a confidence interval for f, we find approximate 90% confidence limits for λ^l and λ^u by solving

$$\frac{932 - \lambda^l}{\sqrt{\lambda^l}} = 1.645 \,, \quad \frac{932 - \lambda^u}{\sqrt{\lambda^u}} = -1.645 \,, \tag{11.5.8}$$

(note that we have made use of the normal approximation to the Poisson distribution). One obtains $\lambda^l = 883.1$ and $\lambda^u = 983.6$, and after division by (11.5.7) the confidence interval turns out to be $0.771 < f < 0.856$.

Confidence limits for the parameter of a Poisson distribution

$\lambda^l\ (w = 0.01)$	$\lambda^l\ (w = 0.05)$	n	$\lambda^u\ (w = 0.05)$	$\lambda^u\ (w = 0.01)$
0.00	0.00	0	3.00	4.61
0.01	0.05	1	4.74	6.64
0.15	0.36	2	6.30	8.41
0.44	0.82	3	7.75	10.05
0.82	1.37	4	9.15	11.60
1.28	1.97	5	10.51	13.11
1.79	2.61	6	11.84	14.57
2.33	3.29	7	13.15	16.00
2.91	3.98	8	14.43	17.40
3.51	4.70	9	15.71	18.78
4.13	5.43	10	16.96	20.14
4.77	6.17	11	18.21	21.49
5.43	6.92	12	19.44	22.82
6.10	7.69	13	20.67	24.14
6.78	8.46	14	21.89	25.45
7.48	9.25	15	23.10	26.74
8.18	10.04	16	24.30	28.03
8.89	10.83	17	25.50	29.31
9.62	11.63	18	26.69	30.58
10.35	12.44	19	27.88	31.85
11.08	13.25	20	29.06	33.10
11.82	14.07	21	30.24	34.36
12.57	14.89	22	31.42	35.60
13.33	15.72	23	32.59	36.84
14.09	16.55	24	33.75	38.08
14.85	17.38	25	34.92	39.31
15.62	18.22	26	36.08	40.53
16.40	19.06	27	37.23	41.76
17.17	19.90	28	38.39	42.98
17.96	20.75	29	39.54	44.19
18.74	21.59	30	40.69	45.40
22.72	25.87	35	46.40	51.41
26.77	30.20	40	52.07	57.35
30.88	34.56	45	57.69	63.23
35.03	38.96	50	63.29	69.07
39.23	43.40	55	68.85	74.86
43.46	47.85	60	74.39	80.62

11.6 The Bayesian Approach

The idea behind the Bayesian method is to view $\mu_{x+\frac{1}{2}}$ as the value assumed by a random variable Θ with prior probability density $u(\vartheta)$. Because of (11.4.2) the posterior density then is

$$\tilde{u}(\vartheta) = \frac{\vartheta^{D_x}\exp(-\vartheta E_x)u(\vartheta)}{\int_0^\infty t^{D_x}\exp(-t\,E_x)u(t)\,dt}. \qquad (11.6.1)$$

The parameter $\mu_{x+\frac{1}{2}}$ may then be estimated by the posterior mean of Θ. The uncertainty attached to the estimate may be quantified by the percentiles of the posterior distribution of Θ.

A common assumption is that the prior distribution of Θ is a gamma distribution with parameters α and β. From (11.6.1) it is easy to see that the posterior distribution will again be a gamma distribution, now with the parameters

$$\tilde{\alpha} = \alpha + D_x, \quad \tilde{\beta} = \beta + E_x. \qquad (11.6.2)$$

Hence we obtain

$$\hat{\mu}_{x+\frac{1}{2}} = \frac{\tilde{\alpha}}{\tilde{\beta}} = \frac{\alpha + D_x}{\beta + E_x} = \frac{\beta}{\beta + E_x}\frac{\alpha}{\beta} + \frac{E_x}{\beta + E_x}\frac{D_x}{E_x}, \qquad (11.6.3)$$

a result that reminds us of credibility theory. An estimator of q_x is obtained by taking the posterior expectation of

$$q_x = 1 - e^{-\Theta}, \qquad (11.6.4)$$

namely

$$\hat{q}_x = 1 - \left(\frac{\tilde{\beta}}{\tilde{\beta}+1}\right)^{\tilde{\alpha}}. \qquad (11.6.5)$$

The percentiles of the posterior gamma distribution can be found using the table of confidence limits of the Poisson parameter, since it can be shown that λ^l is the w-percentile of a gamma distribution with parameters n and 1, and that λ^u is the $(1 - w)$-percentile of a gamma distribution with parameters $n + 1$ and 1. Thus the posterior probability that the true value of Θ lies between $\lambda^l/\tilde{\beta}$ and $\lambda^u/\tilde{\beta}$, is $1 - 2w$. To find λ^l we put $n = \tilde{\alpha}$, and for λ^u we put $n = \tilde{\alpha} - 1$.

11.7 Multiple Causes of Decrement

We return to the model introduced in Chapter 7, where a decrement could be the result of any of m causes. As before we observe the exposure E_x and the number of decrements D_x (for simplicity we shall refer to these as number of deaths). In addition we are informed of the number of deaths by cause j, for $j = 1, 2, \cdots, m$, denoted by $D_{j,x}$. Obviously

$$D_{1,x} + D_{2,x} + \cdots + D_{m,x} = D_x \,. \tag{11.7.1}$$

The probability q_x can be estimated by the methods discussed before. We shall now discuss estimation of the probabilities $q_{j,x}$.

Let us assume piecewise constant forces of decrement, i.e.

$$\mu_{j,x+u} = \mu_{j,x+\frac{1}{2}} \quad \text{for } 0 < u < 1 \,. \tag{11.7.2}$$

Equation (7.2.2) shows then that the aggregate force of decrement also will be piecewise constant. Assuming again that the n lives under observation are independent, we see that the likelihood function is given by

$$\prod_{j=1}^{m} (\mu_{j,x+\frac{1}{2}})^{D_{j,x}} \exp(-\mu_{x+\frac{1}{2}} E_x) \,. \tag{11.7.3}$$

Maximum likelihood estimators are thus

$$\hat{\mu}_{j,x+\frac{1}{2}} = \frac{D_{j,x}}{E_x}, \quad j = 1, 2, \cdots, m \,. \tag{11.7.4}$$

The corresponding estimator for

$$q_{j,x} = \frac{\mu_{j,x+\frac{1}{2}}}{\mu_{x+\frac{1}{2}}} q_x \tag{11.7.5}$$

is then

$$\hat{q}_{j,x} = \frac{D_{j,x}}{D_x} \hat{q}_x \,, \tag{11.7.6}$$

with $\hat{q}_x$ defined by (11.3.7).

In the Bayesian setting the m forces of decrement are considered as realisations of the random variables $\Theta_1, \Theta_2, \cdots, \Theta_m$, which have a prior probability density $u(\vartheta_1, \vartheta_2, \cdots, \vartheta_m)$. The posterior probability density is then proportional to

$$\prod_{j=1}^{m} (\vartheta_j)^{D_{j,x}} \exp(-\vartheta E_x) u(\vartheta_1, \vartheta_2, \cdots, \vartheta_m) \,, \tag{11.7.7}$$

with the definition $\vartheta = \vartheta_1 + \vartheta_2 + \cdots + \vartheta_m$. Now $\hat{\mu}_{j,x+\frac{1}{2}}$ is the posterior mean of Θ_j, and $\hat{q}_{j,x}$ is the posterior mean of

$$\frac{\Theta_j}{\Theta}(1 - e^{-\Theta}) \,, \tag{11.7.8}$$

if we write $\Theta = \Theta_1 + \Theta_2 + \cdots + \Theta_m$.

The analysis is particularly simple under the assumption that the random variables Θ_j are independent, Θ_j having a gamma distribution with parameters α_j and β. In that case the Θ_j are also independent a posteriori, and Θ_j has a gamma distribution with parameters

$$\tilde{\alpha}_j = \alpha_j + D_{j,x}, \quad \tilde{\beta} = \beta + E_x, \tag{11.7.9}$$

which results in the estimate

$$\hat{\mu}_{j,x+\frac{1}{2}} = \frac{\tilde{\alpha}_j}{\tilde{\beta}} = \frac{\alpha_j + D_{j,x}}{\beta + E_x}. \tag{11.7.10}$$

Since the ratio Θ_j/Θ is independent of Θ and has a beta distribution, we can calculate the mean of (11.7.8), obtaining

$$\hat{q}_{j,x+\frac{1}{2}} = \frac{\tilde{\alpha}_j}{\tilde{\alpha}} \hat{q}_x; \tag{11.7.11}$$

here $\tilde{\alpha} = \tilde{\alpha}_1 + \tilde{\alpha}_2 + \cdots + \tilde{\alpha}_m$ and $\hat{q}_x$ is defined by (11.6.5).

11.8 Interpretation of Results

The probability of death at a given age will often be non-stationary in the sense that the general mortality declines as time proceeds. Let us denote the one-year probability of death of a person aged x at calendar time t by q_x^t. On the basis of statistical data from a certain observation period, the values $q_x^t, q_{x+1}^t, q_{x+2}^t, \cdots$ are estimated; here t is taken to be the middle of the observation period. A life table constructed in this way is called a *current*, or *cross-sectional* life table. Such a life table is, of course, an artificial construction.

The probabilities of death and expected values introduced in the preceding chapters all refer to one specific life. Assuming that the initial age of the insured is x at time t, the proper probabilities to use are $q_x^t, q_{x+1}^{t+1}, q_{x+2}^{t+2}, \cdots$. The corresponding life table is called a *longitudinal* or *generation* life table, since it relates to the generation of persons born at time $t - x$. This life table defines the probability distribution of $K = K(x)$. The probabilities of death in a generation life table must be estimated by a suitable method of extrapolation.

Appendix A. Commutation Functions

A.1 Introduction

In this appendix we give an introduction to the use of commutation functions. These functions were invented in the 18th century and achieved great popularity, which can be ascribed to two reasons:

Reason 1

Tables of commutation functions simplify the calculation of numerical values for many actuarial functions.

Reason 2

Expected values such as net single premiums may be derived within a deterministic model closely related to commutation functions.

Both reasons have lost their significance, the first with the advent of powerful computers, the second with the growing acceptance of models based on probability theory, which allows a more complete understanding of the essentials of insurance. It may therefore be taken for granted that the days of glory for the commutation functions now belong to the past.

A.2 The Deterministic Model

Imagine a cohort of lives, all of the same age, observed over time, and denote by l_x the number still living at age x. Thus $d_x = l_x - l_{x+1}$ is the number of deaths between the ages of x and $x + 1$.

Probabilities and expected values may now be derived from simple proportions and averages. So is, for instance,

$$_t p_x = l_{x+t}/l_x \tag{A.2.1}$$

the proportion of persons alive at age $x + t$, relative to the number of persons alive at age x, and the probability that a life aged x will die within a year is

$$q_x = d_x/l_x \,. \tag{A.2.2}$$

In Chapter 2 we introduced the expected curtate future lifetime of a life aged x. Replacing $_kp_x$ by l_{x+k}/l_x in (2.4.3), we obtain

$$e_x = \frac{l_{x+1} + l_{x+2} + \cdots}{l_x} \,. \tag{A.2.3}$$

The numerator in this expression is the total number of complete future years to be "lived" by the l_x lives (x), so that e_x is the average number of completed years left.

A.3 Life Annuities

We first consider a life annuity-due with annual payments of 1 unit, as introduced in Section 4.2, the net single premium of which annuity was denoted by $\ddot{a}_x$. Replacing $_kp_x$ in (4.2.5) by l_{x+k}/l_x, we obtain

$$\ddot{a}_x = \frac{l_x + vl_{x+1} + v^2l_{x+2} + \cdots}{l_x} \,, \tag{A.3.1}$$

or

$$l_x \,\ddot{a}_x = l_x + vl_{x+1} + v^2l_{x+2} + \cdots \,. \tag{A.3.2}$$

This result is often referred to as the *equivalence principle*, and its interpretation within the deterministic model is evident: if each of the l_x persons living at age x were to buy an annuity of the given type, the sum of net single premiums (the left hand side of (A.3.2)) would equal the present value of the benefits (the right hand side of (A.3.2)).

Multiplying both numerator and denominator in (A.3.1) by v^x, we find

$$\ddot{a}_x = \frac{v^xl_x + v^{x+1}l_{x+1} + v^{x+2}l_{x+2} + \cdots}{v^xl_x} \,. \tag{A.3.3}$$

With the abbreviations

$$D_x = v^xl_x \,, \quad N_x = D_x + D_{x+1} + D_{x+2} + \cdots \tag{A.3.4}$$

we then obtain the simple formula

$$\ddot{a}_x = \frac{N_x}{D_x} \,. \tag{A.3.5}$$

Thus the manual calculation of $\ddot{a}_x$ is extremely easy if tables of the commutation functions D_x and N_x are available. The function D_x is called the *"discounted number of survivors"*.

Similarly one may obtain formulas for the net single premium of a temporary life annuity,

$$\ddot{a}_{x:\overline{n}|} = \frac{N_x - N_{x+n}}{D_x}, \tag{A.3.6}$$

immediate life annuities,

$$a_x = \frac{N_{x+1}}{D_x}, \tag{A.3.7}$$

and general annuities with annual payments: formula (4.4.2) may naturally be translated to

$$E(Y) = \frac{r_0 D_x + r_1 D_{x+1} + r_2 D_{x+2} + \cdots}{D_x}. \tag{A.3.8}$$

For the special case $r_k = k + 1$ we obtain the formula

$$(I\ddot{a})_x = \frac{S_x}{D_x}; \tag{A.3.9}$$

here the commutation function S_x is defined by

$$\begin{aligned} S_x &= D_x + 2D_{x+1} + 3D_{x+2} + \cdots \\ &= N_x + N_{x+1} + N_{x+2} + \cdots. \end{aligned} \tag{A.3.10}$$

A.4 Life Insurance

In addition to (A.3.4) and (A.3.10) we now define the commutation functions

$$\begin{aligned} C_x &= v^{x+1} d_x, \\ M_x &= C_x + C_{x+1} + C_{x+2} + \cdots, \\ R_x &= C_x + 2C_{x+1} + 3C_{x+2} + \cdots \\ &= M_x + M_{x+1} + M_{x+2} + \cdots. \end{aligned} \tag{A.4.1}$$

Replacing $_k p_x q_{x+k}$ in equation (3.2.3) by d_{x+k}/l_x, we obtain

$$\begin{aligned} A_x &= \frac{v d_x + v^2 d_{x+1} + v^3 d_{x+2} + \cdots}{l_x} \\ &= \frac{C_x + C_{x+1} + C_{x+2} + \cdots}{D_x} \\ &= \frac{M_x}{D_x}. \end{aligned} \tag{A.4.2}$$

Similarly one obtains

$$\begin{aligned} (IA)_x &= \frac{v d_x + 2v^2 d_{x+1} + 3v^3 d_{x+2} + \cdots}{l_x} \\ &= \frac{C_x + 2C_{x+1} + 3C_{x+2} + \cdots}{D_x} \\ &= \frac{R_x}{D_x}. \end{aligned} \tag{A.4.3}$$

Obviously these formulae may be derived within the deterministic model by means of the equivalence principle. In order to determine A_x one would start with

$$l_x A_x = v d_x + v^2 d_{x+1} + v^3 d_{x+2} + \cdots \qquad (A.4.4)$$

by imagining that l_x persons buy a whole life insurance of 1 unit each, payable at the end of the year of death, in return for a net single premium.

Corresponding formulae for term and endowment insurances are

$$
\begin{aligned}
A^1_{x:\overline{n}|} &= \frac{M_x - M_{x+n}}{D_x}, \\
A_{x:\overline{n}|} &= \frac{M_x - M_{x+n} + D_{x+n}}{D_x}, \\
(IA)^1_{x:\overline{n}|} &= \frac{C_x + 2C_{x+1} + 3C_{x+2} + \cdots + nC_{x+n-1}}{D_x} \\
&= \frac{M_x + M_{x+1} + M_{x+2} + \cdots + M_{x+n-1} - nM_{x+n}}{D_x} \\
&= \frac{R_x - R_{x+n} - nM_{x+n}}{D_x}, \qquad (A.4.5)
\end{aligned}
$$

which speak for themselves.

The commutation functions defined in (A.4.1) can be expressed in terms of the commutation functions defined in Section 3. From $d_x = l_x - l_{x+1}$ follows

$$C_x = v D_x - D_{x+1}. \qquad (A.4.6)$$

Summation yields the identities

$$M_x = v N_x - (N_x - D_x) = D_x - d N_x \qquad (A.4.7)$$

and

$$R_x = N_x - d S_x. \qquad (A.4.8)$$

Dividing both equations by D_x, we retrieve the identities

$$
\begin{aligned}
A_x &= 1 - d\,\ddot{a}_x, \\
(IA)_x &= \ddot{a}_x - d(I\ddot{a})_x, \qquad (A.4.9)
\end{aligned}
$$

see equations (4.2.8) and (4.5.2).

A.5 Net Annual Premiums and Premium Reserves

Consider a whole life insurance with 1 unit payable at the end of the year of death, and payable by net annual premiums. Using (A.3.5) and (A.4.2) we find

$$P_x = \frac{A_x}{\ddot{a}_x} = \frac{M_x}{N_x}. \qquad (A.5.1)$$

Of course, the deterministic approach, i.e. the condition

$$P_x l_x + v P_x l_{x+1} + v^2 P_x l_{x+2} + \cdots = v d_x + v^2 d_{x+1} + v^3 d_{x+2} + \cdots \qquad (A.5.2)$$

leads to the same result.

The net premium reserve at the end of year k then becomes

$$_k V_x = A_{x+k} - P_x \ddot{a}_{x+k} = \frac{M_{x+k} - P_x N_{x+k}}{D_{x+k}} . \qquad (A.5.3)$$

This result may also be obtained by the deterministic condition

$$\begin{aligned}
_k V_x l_{x+k} &+ P_x l_{x+k} + v P_x l_{x+k+1} + v^2 P_x l_{x+k+2} + \cdots \\
&= v d_{x+k} + v^2 d_{x+k+1} + v^3 d_{x+k+2} + \cdots .
\end{aligned} \qquad (A.5.4)$$

Here one imagines that each person alive at time k is allotted the amount $_k V_x$; the condition (A.5.4) states that the sum of the net premium reserve and the present value of future premiums must equal the present value of all future benefit payments.

The interested reader should be able to apply this technique to other, more general situations.

Appendix B. Simple Interest

In practice, the accumulation factor for a time interval of length h is occasionally approximated by

$$(1 + i)^h \approx 1 + hi \, . \tag{B.1}$$

This approximation is obtained by neglecting all but the linear terms in the Taylor expansion of the left hand side above; alternatively the right hand side may be obtained by linear interpolation between $h = 0$ and $h = 1$. Similarly an approximation for the discount factor for an interval of length h is

$$v^h = (1 - d)^h \approx 1 - hd \, . \tag{B.2}$$

The approximations (B.1) and (B.2) have little practical importance since the advent of pocket calculators.

Interest on transactions with a savings account is sometimes calculated according to the following rule: If an amount of r is deposited (drawn) at time u ($0 < u < 1$), it is valued at time 0 as

$$rv^u \approx r(1 - ud) \, . \tag{B.3}$$

At the end of the year (time 1) the amount is valued as

$$
\begin{aligned}
r(1 + i)^{1-u} &= r(1 + i)v^u \approx r(1 + i)(1 - ud) \\
&= r\{1 + (1 - u)i\} \, .
\end{aligned} \tag{B.4}
$$

This technique amounts to accumulation from time u to time 1 according to (B.1) or discounting from u to 0 according to (B.2). With a suitably chosen variable force of interest the rule is exact; this variable force of interest is determined by equating the accumulation factors:

$$1 + (1 - u)i = \exp\left(\int_u^1 \delta(t)dt \right) \, . \tag{B.5}$$

Differentiating the logarithms gives the expression

$$\delta(u) = \frac{i}{1 + (1 - u)i} = \frac{d}{1 - ud} \tag{B.6}$$

for $0 < u < 1$. The force of interest thus increases from $\delta(0) = d$ to $\delta(1) = i$ during the year.

The technique sketched above is based on the assumption that the accumulation factor for the time interval from u to 1 is a linear function of u; this assumption is analogous to *Assumption c* of Section 2.6, concerning mortality for fractional durations. The similarity between (B.6) and (2.6.10) is evident.

Appendix C

Exercises

C.0   Introduction

These exercises provide two types of practice. The first type consists of theoretical exercises, some demonstrations, and manipulation of symbols. Some of these problems of the first kind are based on Society of Actuaries questions from examinations prior to May 1990. The second type of practice involves using a spreadsheet program. Many exercises are solved in Appendix D. For the spreadsheet exercises, we give a guide to follow in writing your own program. For the theoretical exercises, we usually give a complete description. We provide guides for solving the spreadsheet problems, rather than computer codes. The student should write a program and use the guide to verify it. We use the terminology of *Excel* in the guides. The terminology of other programs is analogous.

I would like to thank Hans Gerber for allowing me to contribute these exercises to his textbook. It is a pleasure to acknowledge the assistance of Georgia State University graduate students, Masa Ozeki and Javier Suarez who helped by checking solutions and proofreading the exercises.

I hope that students will find these exercises challenging and enlightening.

Atlanta, June 1995   *Samuel H. Cox*

C.1 Mathematics of Compound Interest: Exercises

A *bond* is a contract obligating one party, the borrower or bond issuer, to pay to the other party, the lender or bondholder, a series of future payments defined by the face value, F, and the coupon rate, c. At the end of each future period the borrower pays cF to the lender. The bond matures after N periods with a final coupon payment and a simultaneous payment of the redemption value C. Usually C is equal to F. Investors (lenders) require a yield to maturity of $i \geq 0$ effective per period. The price, P, is the present value of future cash flows paid to the bondholder. The five values are related by the following equation.

$$P = cF\frac{1 - v^N}{i} + Cv^N$$

where $v = 1/(1 + i)$.

C.1.1 Theory Exercises

1. Show that
$$i^{(m)} - d^{(m)} = \frac{i^{(m)}d^{(m)}}{m}.$$

2. Show that $d < d^{(2)} < d^{(3)} < \ldots < \delta < \ldots < i^{(3)} < i^{(2)} < i$ and
$$i^{(m)} - d^{(n)} \leq \frac{i^2}{min(m, n)}.$$

3. A company must retire a bond issue with five annual payments of 15,000. The first payment is due on December 31, 1999. In order to accumulate the funds, the company begins making annual payments of X on January 1, 1990 into an account paying effective annual interest of 6%. The last payment is to be made on January 1, 1999. Calculate X.

4. At a nominal annual rate of interest j, convertible semiannually, the present value of a series of payments of 1 at the end of every 2 years, which continue forever, is 5.89. Calculate j.

5. A perpetuity consists of yearly increasing payments of $(1 + k)$, $(1 + k)^2$, $(1 + k)^3$, etc., commencing at the end of the first year. At an annual effective interest rate of 4%, the present value one year before the first payment is 51. Determine k.

6. Six months before the first coupon is due a ten-year semi-annual coupon bond sells for 94 per 100 of face value. The rate of payment of coupons is 10% per year. The yield to maturity for a zero-coupon ten-year bond is 12%. Calculate the yield to maturity of the coupon payments.

7. A loan of 1000 at a nominal rate of 12% convertible monthly is to be repaid by six monthly payments with the first payment due at the end of one month. The first three payments are x each, and the final three payments are $3x$ each. Calculate x.

8. A loan of 4000 is being repaid by a 30-year increasing annuity immediate. The initial payment is k, each subsequent payment is k larger than the preceding payment. The annual effective interest rate is 4%. Calculate the principal outstanding immediately after the ninth payment.

9. John pays 98.51 for a bond that is due to mature for 100 in one year. It has coupons at 4% convertible semiannually. Calculate the annual yield rate convertible semiannually.

10. The death benefit on a life insurance policy can be paid in four ways. All have the same present value:

(i) A perpetuity of 120 at the end of each month, first payment one month after the moment of death;

(ii) Payments of 365.47 at the end of each month for n years, first payment one month after the moment of death;

(iii) A payment of 17,866.32 at the end of n years after the moment of death; and

(iv) A payment of X at the moment of death.

Calculate X.

C.1.2   Spreadsheet Exercises

1. A serial bond with a face amount of 1000 is priced at 1145. The owner of the bond receives annual coupons of 12% of the outstanding principal. The principal is repaid by the following schedule:

(i) 100 at the end of each years 10 through 14, and

(ii) 500 at the end of year 15.

(a) Calculate the investment yield using the built-in Goal Seek procedure.

(b) Use the graphic capability of the spreadsheet to illustrate the investment yield graphically. To do this, construct a Data Table showing various investment yield values and the corresponding bond prices. From the graph, determine which yield corresponds to a price of 1,145.

2. A deposit of 100,000 is made into a newly established fund. The fund pays nominal interest of 12% convertible quarterly. At the end of each six months a withdrawal is made from the fund. The first withdrawal is X, the second is $2X$, the third is $3X$, and so on. The last is the sixth withdrawal which exactly exhausts the fund. Calculate X.

3. A loan is to be repaid by annual installments 100, 200, 300, 300, 200 and 100. In the fifth installment, the amount of principal repayment is equal to six times the amount of interest. Calculate the annual effective interest rate.

4. A company borrows 10,000. Interest of 350 is paid semiannually, but no principal is paid until the entire loan is repaid at the end of 5 years. In order to accumulate the principal of the loan at the end of five years, the company makes equal semiannual deposits, the first due in six months, into a fund that credits interest at a nominal annual rate of 6% compounded semiannually. Calculate the internal rate of return effective per year for the company on the entire transaction.

5. Deposits of 100 are made into a fund at the beginning of each year for 10 years. Beginning ten years after the last deposit, X is withdrawn each year from the fund in perpetuity.
(a) $i = 10\%$. Calculate X.
(b) Draw the graph of X as a function of i for i varying from 1% to 21% in increments of 2%.

6. A bank credits savings accounts with 8% annual effective interest on the first 100,000 of beginning year account value and 9% on the excess over 100,000. An initial deposit of 300,000 is made. One year later level annual withdrawals of X begin and run until the account is exactly exhausted with the tenth withdrawal. Calculate X.

7. In order to settle a wrongful injury claim, an annuity is purchased from an insurance company. According to the annuity contract, the insurer is obliged to make the following future payments on July 1 of each year indicated:

Year	Amount
1995	50,000
1996	60,000
1997	75,000
1998	100,000
1999	125,000
2000	200,000

The insurer is considering hedging its future liability under the annuity contract by purchasing government bonds. The financial press publishes the market prices for the following government bonds available for sale on July 1, 1994. Each bond has a face amount of 10,000, each pays annual coupons on July 1, and the first coupon payment is due in one year.

Maturity	Coupon Rate	Price
1995	4.250%	9,870
1996	7.875%	10,180
1997	5.500%	9,600
1998	5.250%	9,210
1999	6.875%	9,740
2000	7.875%	10,120

Determine how many bonds of each maturity the insurer should buy on July 1, 1994 so that the aggregate cash flow from the bonds will exactly match the insurer's obligation under the terms of the claim settlement. Assume that fractions of bonds may be purchased.

8. A loan of 100,000 is repayable over 20 years by semiannual payments of 2500, plus 5% interest (per year convertible twice per year) on the outstanding balance. Immediately after the tenth payment the lender sells the loan for 65,000. Calculate the corresponding market yield to maturity of the loan (per year convertible twice per year).

9. A bond with face value 1000 has 9% annual coupons. The borrower may call the bond at the end of years 10 though 15 by paying the face amount plus a call premium, according to the schedule:

Year	10	11	12	13	14	15
Premium	100	80	60	40	20	0

For example, if the borrower elects to repay the debt at the end of year 11 (11 years from now), a payment of $1000 + 80 = 1080$ plus the coupon then due of 90 would be paid to the lender. The debt is paid; no further payments would be made. Calculate the price now, one year before the next coupon payment, to be certain of a yield of at least 8% to the call date.

10. Equal deposits of 200 are made to a bank account at the beginning of each quarter of a year for five years. The bank pays interest from the date of deposit at an annual effective rate of i. One quarter year after the last deposit the account balance is 5000. Calculate i.

C.2 The Future Lifetime of a Life Aged x: Exercises

These exercises sometimes use the commutation function notation introduced in Appendix A and the following notation with regard to mortality tables. The Illustrative Life Table is given in Appendix E. It is required for some exercises.

A mortality table covering the range of ages x $(0 \le x < \omega)$ is denoted by l_x, which represents the number l_0 of the new-born lives who survive to age x. The probability of surviving to age x is $s(x) = l_x/l_0$. The rule for calculating conditional probabilities establishes this relationship to $_tp_x$:

$$_tp_x = \Pr(T(0) > x + t | T(0) > x) = \frac{s(x+t)}{s(x)} = \frac{l_{x+t}}{l_x}.$$

In the case that the conditioning involves more information than mere survival, the notation $_tp_{[x]}$ is used. Thus if a person age x applies for insurance and is found to be in good health, the mortality function is denoted $_tp_{[x]}$ rather than $_tp_x$. The notation $[x]$ tells us that some information in addition to $T(0) > x$ was used in preparing the survival distribution. This gives rise to the select and ultimate mortality table discussed in the text.

Here are some additional mortality functions:

$$
\begin{aligned}
m_x &= \quad \text{central death rate} = \frac{d_x}{L_x} \\
L_x &= \quad \text{average number of survivors to } (x, x+1) \\
&= \quad \int_x^{x+1} l_y\,dy = \int_0^1 l_{x+t}\,dt \\
d_x &= \quad \text{number of deaths in } (x, x+1) = l_x - l_{x+1}.
\end{aligned}
$$

Since $_tp_x\mu_{x+t} = -\frac{d}{dt}\,_tp_x$, then in terms of l_x we have $l_{x+t}\mu_{x+t} = -\frac{d}{dt}l_{x+t}$ or, letting $y = x + t$, we have $l_y\mu_y = -\frac{d}{dy}l_y$ for all y. The following are useful for calculating $\mathrm{Var}(T)$ and $\mathrm{Var}(K)$:

$$
\begin{aligned}
\mathrm{E}[T^2] &= \int_0^\infty t^2\,_tp_x\mu_{x+t}\,dt \\
&= \int_0^\infty 2t\,_tp_x\,dt \\
\mathrm{E}[K^2] &= \sum_{k=1}^\infty k^2\,_{k-1}p_x q_{x+k-1} \\
&= \sum_{k=0}^\infty (2k+1)\,_{k+1}p_x.
\end{aligned}
$$

C.2.1   Theory Exercises

1. Given:
$$_tp_x = \frac{100 - x - t}{100 - x}$$
for $0 \le x < 100$ and $0 \le t \le 100 - x$. Calculate μ_{45}.

2. Given:
$$_tp_x = 1 - \left(\frac{t}{100}\right)^{1.5}$$
for $x = 60$ and $0 < t < 100$. Calculate $E[T(x)]$.

3. Given: $\mu_{x+t} = \dfrac{1}{85 - t} + \dfrac{3}{105 - t}$ for $0 \le t < 85$. Calculate $_{20}p_x$.

4. Given: $_tp_x = \left(\dfrac{1 + x}{1 + x + t}\right)^3$ for $t \ge 0$. Calculate the complete life expectancy of a person age $x = 41$.

5. Given: $q_x = 0.200$. Calculate $m_x = \dfrac{q_x}{\int_0^1 {}_tp_x\, dt}$ using *assumption c*, the Balducci assumption.

6. Given:

(i) μ_{x+t} is constant for $0 \le t < 1$ and

(ii) $q_x = 0.16$.

Calculate the value of t for which $_tp_x = 0.95$.

7. Given:

(i) The curve of death $l_x\mu_x$ is constant for $0 \le x < \omega$.

(ii) $\omega = 100$.

Calculate the variance of the remaining lifetime random variable $T(x)$ at $x = 88$.

8. Given:

(i) When the force of mortality is μ_{x+t}, $0 < t < 1$, then $q_x = 0.05$.

(ii) When the force of mortality is $\mu_{x+t} - c$, $0 < t < 1$, then $q_x = 0.07$.

Calculate c.

9. Prove:

(i) $_tp_x = \exp\left(-\int_x^{x+t} \mu_s\, ds\right)$ and

(ii) $\frac{\partial}{\partial x}{}_tp_x = (\mu_x - \mu_{x+t})_tp_x$.

10. You are given the following excerpt from a select and ultimate mortality table with a two-year select period.

x	$100q_{[x]}$	$100q_{[x]+1}$	$100q_{x+2}$
30	0.438	0.574	0.699
31	0.453	0.599	0.734
32	0.472	0.634	0.790
33	0.510	0.680	0.856
34	0.551	0.737	0.937

Calculate $100(_{1|}q_{[30]+1})$.

11. Given:

$$l_x = (121 - x)^{1/2}$$

for $0 \leq x \leq 121$. Calculate the probability that a life age 21 will die after attaining age 40, but before attaining age 57.

12. Given the following table of values of e_x:

Age x	e_x
75	10.5
76	10.0
77	9.5

Calculate the probability that a life age 75 will survive to age 77. Hint: Use the recursion relation $e_x = p_x(1 + e_{x+1})$.

13. Mortality follows de Moivre's law and $E[T(16)] = 36$. Calculate $Var(T(16))$.

14. Given:

$$_tp_{30} = \frac{7800 - 70t - t^2}{7800}$$

for $0 \leq t \leq 60$. Calculate the exact value of $q_{50} - \mu_{50}$.

15. Given:

$$_tp_x = \left(\frac{100 - x - t}{100 - x}\right)^2$$

for $0 \leq t \leq 100 - x$. Calculate $Var(T(x))$.

16. Given: $q_x = 0.420$ and *assumption b* applies to the year of age x to $x + 1$. Calculate m_x, the central death rate exactly. (See exercise 5.)

17. Consider two independent lives, which are identical except that one is a smoker and the other is a non-smoker. Given:

(i) μ_x is the force of mortality for non-smokers for $0 \leq x < \omega$.

(ii) $c\mu_x$ is the force of mortality for smokers for $0 \leq x < \omega$, where c is a constant, $c > 1$.

Calculate the probability that the remaining lifetime of the smoker exceeds that of the non-smoker.

18. Derive an expression for the derivative of q_x with respect to x in terms of the force of mortality.

19. Given: $\mu_x = kx$ for all $x > 0$ where k is a positive constant and $_{10}p_{35} = 0.81$. Calculate $_{20}p_{40}$.

20. Given:

(i) $l_x = 1000(\omega^3 - x^3)$ for $0 \le x \le \omega$ and

(ii) $E[T(0)] = 3\omega/4$.

Calculate $\text{Var}(T(0))$.

C.2.2   Spreadsheet Exercises

1. Put the Illustrative Life Table l_x values into a spreadsheet. Calculate d_x and $1000q_x$ for $x = 0, 1, \ldots, 99$.

2. Calculate $e_x, x = 0, 1, 2, \ldots, 99$ for the Illustrative Life Table. Hint: Use formula (2.4.3) to get $e_{99} = p_{99} = 0$ for this table. The recursive formula $e_x = p_x(1 + e_{x+1})$ follows from (2.4.3). Use it to calculate from the higher age to the lower.

3. A sub-standard mortality table is obtained from a standard table by adding a constant c to the force of mortality. This results in sub-standard mortality rates q_x^s which are related to the standard rates q_x by $q_x^s = 1 - e^{-c}(1 - q_x)$. Use the Illustrative Life Table for the standard mortality. A physician examines a life age $x = 40$ and determines that the expectation of remaining lifetime is 10 years. Determine the constant c, and the resulting substandard table. Prepare a table and graph of the mortality ratio (sub-standard q_x^s to standard q_x) by year of age, beginning at age 40.

4. Draw the graph of $\mu_x = Bc^x, x = 0, 1, 2, \ldots, 110$ for $B = 0.0001$ and each value of $c = 1.01, 1.05, 1.10, 1.20$. Calculate the corresponding values of l_x and draw the graphs. Use $l_0 = 100,000$ and round to an integer.

5. Let $q_x = 0.10$. Draw the graphs of μ_{x+u} for u running from 0 to 1 increments of 0.05 for each of the interpolation formulas given by *assumptions a, b,* and *c*.

6. Substitute $_uq_x$ for μ_{x+u} in Exercise 5 and rework.

7. Use the method of least squares (and the spreadsheet Solver feature) to fit a Gompertz distribution to the Illustrative Life Table values of $_tp_x$ for $x = 50$ and $t = 1, 2, \ldots, 50$. Draw the graph of the table values and the Gompertz values on the same axes.

8. A sub-standard mortality table is obtained from a standard table by multiplying the standard q_x by a constant $k \ge 1$, subject to an upper bound of 1.

Thus the substandard q_x^s mortality rates are related to the standard rates q_x by $q_x^s = \min(kq_x, 1)$.

(a) For values of k ranging from 1 to 10 in increments of 0.5, calculate points on the graph of $_tp_x^s$ for age $x = 45$ and t running from 0 to the end of the table in increments of one year. Draw the graphs in a single chart.

(b) Calculate the sub-standard life expectancy at age $x = 45$ for each value of k in (a).

C.3   Life Insurance

C.3.1   Theory Exercises

1. Given:

(i) The survival function is $s(x) = 1 - x/100$ for $0 \leq x \leq 100$.

(ii) The force of interest is $\delta = 0.10$.

Calculate $50{,}000 \bar{A}_{30}$.

2. Show that

$$\frac{(IA)_x - A^1_{x:\overline{1}|}}{(IA)_{x+1} + A_{x+1}}$$

simplifies to vp_x.

3. Z_1 is the present value random variable for an n-year continuous endowment insurance of 1 issued to (x). Z_2 is the present value random variable for an n-year continuous term insurance of 1 issued to x. Given:

(i) $\text{Var}(Z_2) = 0.01$

(ii) $v^n = 0.30$

(iii) $_np_x = 0.8$

(iv) $\text{E}[Z_2] = 0.04$.

Calculate $\text{Var}(Z_1)$.

4. Use the Illustrative Life Table and $i = 5\%$ to calculate $A_{45:\overline{20}|}$.

5. Given:

(i) $A_{x:\overline{n}|} = u$

(ii) $A^1_{x:\overline{n}|} = y$

(iii) $A_{x+n} = z$.

Determine the value of A_x in terms of u, y, and z.

6. A continuous whole life insurance is issued to (50). Given:

(i) Mortality follows de Moivre's law with $\omega = 100$.

(ii) Simple interest with $i = 0.01$.

(iii) $b_t = 1000 - 0.1t^2$.

Calculate the expected value of the present value random variable for this insurance.

7. Assume that the forces of mortality and interest are each constant and denoted by μ and δ, respectively. Determine $\text{Var}(v^T)$ in terms of μ and δ.

8. For a select and ultimate mortality table with a one-year select period, $q_{[x]} = 0.5q_x$ for all $x \geq 0$. Show that $A_x - A_{[x]} = 0.5vq_x(1 - A_{x+1})$.

9. A single premium whole life insurance issued to (x) provides 10,000 of insurance during the first 20 years and 20,000 of insurance thereafter, plus a return without interest of the net single premium if the insured dies during the first 20 years. The net single premium is paid at the beginning of the first year. The death benefit is paid at the end of the year of death. Express the net single premium using commutation functions.

10. A ten-year term insurance policy issued to (x) provides the following death benefits payable at the end of the year of death.

Year of Death	Death Benefit
1	10
2	10
3	9
4	9
5	9
6	8
7	8
8	8
9	8
10	7

Express the net single premium for this policy using commutation functions.

11. Given:

(i) The survival function is $s(x) = 1 - x/100$ for $0 \leq x \leq 100$.

(ii) The force of interest is $\delta = 0.10$.

(iii) The death benefit is paid at the moment of death.

Calculate the net single premium for a 10-year endowment insurance of 50,000 for a person age $x = 50$.

12. Given:

(i) $s(x) = e^{-0.02x}$ for $x \geq 0$

(ii) $\delta = 0.04$.

Calculate the median of the present value random variable $Z = v^T$ for a whole life policy issued to (y).

13. A 2-year term insurance policy issued to (x) pays a death benefit of 1 at the end of the year of death. Given:

(i) $q_x = 0.50$

(ii) $i = 0$

(iii) $\text{Var}(Z) = 0.1771$

where Z is the present value of future benefits. Calculate q_{x+1}.

14. A 3-year term life insurance to (x) is defined by the following table:

Year t	Death Benefit	q_{x+t}
0	3	0.20
1	2	0.25
2	1	0.50

Given: $v = 0.9$, the death benefits are payable at the end of the year of death and the expected present value of the death benefit is Π. Calculate the probability that the present value of the benefit payment that is actually made will exceed Π.

15. Given:

(i) $A_{76} = 0.800$

(ii) $D_{76} = 400$

(iii) $D_{77} = 360$

(iv)] $i = 0.03$.

Calculate A_{77} by use of the recursion formula (3.6.1).

16. A whole life insurance of 50 is issued to (x). The benefit is payable at the moment of death. The probability density function of the future lifetime, T, is

$$g(t) = \begin{cases} t/5000 & \text{for } 0 \leq t \leq 100 \\ 0 & \text{elsewhere.} \end{cases}$$

The force of interest is constant: $\delta = 0.10$. Calculate the net single premium.

17. For a continuous whole life insurance, $E[v^{2T}] = 0.25$. Assume the forces of mortality and interest are each constant. Calculate $E[v^T]$.

18. There are 100 club members age x who each contribute an amount w to a fund. The fund earns interest at $i = 10\%$ per year. The fund is obligated to pay 1000 at the moment of death of each member. The probability is 0.95 that the fund will meet its benefit obligations. Given the following values calculated at $i = 10\%$: $\bar{A}_x = 0.06$ and $^2\bar{A}_x = 0.01$. Calculate w. Assume that the future lifetimes are independent and that a normal distribution may be used.

19. An insurance is issued to (x) that

(i) pays 10,000 at the end of 20 years if x is alive and

(ii) returns the net single premium Π at the end of the year of death if (x) dies during the first 20 years.

Express II using commutation functions.

20. A whole life insurance policy issued to (x) provides the following death benefits payable at the end of the year of death.

Year of Death	Death Benefit
1	10
2	10
3	9
4	9
5	9
6	8
7	8
8	8
9	8
10	7
each other year	7

Calculate the net single premium for this policy.

C.3.2 Spreadsheet Exercises

1. Calculate the A_x column of the Illustrative Life Table at $i = 5\%$. Use the recursive method suggested by formula (3.6.1). Construct a graph showing the values of A_x for $i = 0, 2.5\%, 5\%, 7.5\%, 10\%$ and $x = 0, 1, 2, \ldots, 99$.

2. The formula for increasing life insurances, in analogy to (3.6.1), is $(IA)_x = vq_x + vp_x(A_{x+1} + IA_{x+1})$. Use this (and (3.6.1)) to calculate a table of values of $(IA)_x$ for the Illustrative Life Table and $i = 5\%$.

3. Calculate the net single premium of an increasing 20 year term insurance for issue age $x = 25$, assuming that the benefit is 1 the first year, $1 + g$ the second year, $(1 + g)^2$ the third year and so on. Use the Illustrative Life Table at $i = 5\%$ and $g = 6\%$. Try to generalize to a table of premiums for all issue ages $x = 0, 1, 2, \ldots, 99$.

4. Calculate the net single premium for a decreasing whole life insurance with an initial benefit of $100 - x$ at age x, decreasing by 1 per year. The benefit is paid at the moment of death. Use the Illustrative Life Table at $i = 5\%$ and $x = 50$. Generalize so that x and i are input cell values, and your spreadsheet calculates the premium for reasonable interest rates and ages.

5. For a life age $x = 35$, calculate the variance of the present value random variable for a whole life insurance of 1000. The interest rate i varies from 0 to 25% by increments of 0.5%. Mortality follows the Illustrative Life Table. Draw the graph.

C.4   Life Annuities

C.4.1   Theory Exercises

1. Using *assumption a* and the Illustrative Life Table with interest at the effective annual rate of 5%, calculate $\ddot{a}^{(2)}_{40:\overline{30}|}$.

2. Demonstrate that

$$\frac{(I\ddot{a})_x - \ddot{a}_{x:\overline{1}|}}{(I\ddot{a})_{x+1} + \ddot{a}_{x+1}}$$

simplifies to $a_{x:\overline{1}|}$.

3. $\left(\bar{I}_{\overline{n}|}\bar{a}\right)_x$ is equal to $E[Y]$ where

$$Y = \begin{cases} (\bar{I}\bar{a})_{\overline{T}|} & \text{if } 0 \le T < n \text{ and} \\ (\bar{I}\bar{a})_{\overline{n}|} + n\left({}_{n|}\bar{a}_{\overline{T-n}|}\right) & \text{if } T \ge n \end{cases}$$

The force of mortality is constant, $\mu_x = 0.04$ for all x, and the force of interest is constant, $\delta = 0.06$. Calculate $\frac{\partial}{\partial n}\left(\bar{I}_{\overline{n}|}\bar{a}\right)_x$.

4. Given the following information for a 3-year temporary life annuity due, contingent on the life of (x):

t	Payment	p_{x+t}
0	2	0.80
1	3	0.75
2	4	0.50

and $v = 0.9$. Calculate the variance of the present value of the indicated payments.

5. Given:

(i) $l_x = 100{,}000(100 - x)$, $0 \le x \le 100$ and

(ii) $i = 0$.

Calculate $(I\bar{a})_{95}$ exactly.

6. Calculate ${}_{10|}\ddot{a}^{(12)}_{25:\overline{10}|}$ using the Illustrative Life Table, *assumption a* and $i = 5\%$. (The symbol denotes an annuity issued on a life age 25, the first payment deferred 10 years, paid in level monthly payments at a rate of 1 per year during the lifetime of the annuitant but not more than 10 years.)

7. Given:

(i)

x	69	70	71	72	$\cdots$	79	80	81	82
S_x	77,938	67,117	57,520	49,043	$\cdots$	13,483	10,875	8,691	6,875

(ii) $\alpha(12) = 1.00028$ and $\beta(12) = 0.46812$

(iii) *assumption a* applies: deaths are distributed uniformly over each year of age.

Calculate $(I\ddot{a})^{(12)}_{70:\overline{10}|}$.

8. Show:

$$_np_x d\ddot{a}_{\overline{n}|} + \sum_{k=0}^{n-1}(1 - v^{k+1})_kp_xq_{x+k} = 1 - A_{x:\overline{n}|}$$

9. Y is the present value random variable of a whole life annuity due of 1 per year issued to (x). Given: $\ddot{a}_x = 10$, evaluated with $i = 1/24 = e^\delta - 1$, and $\ddot{a}_x = 6$, evaluated with $i = e^{2\delta} - 1$. Calculate the variance of Y.

10. $\ddot{a}_{x:\overline{n}|}$ is equal to E[Y] where

$$Y = \begin{cases} \ddot{a}_{\overline{K+1}|} & \text{if } 0 \le K < n \text{ and} \\ \ddot{a}_{\overline{n}|} & \text{if } K \ge n. \end{cases}$$

Show that

$$\text{Var}[Y] = \frac{M(-2\delta) - M(-\delta)^2}{d^2}$$

where $M(u) = E(e^{u\,\min(K+1,n)})$ is the moment generating function of the random variable $\min(K+1,n)$.

11. Given $i = 0.03$ and commutation function values:

x	27	28	29	30	31
S_x	1,868	1,767	1,670	1,577	1,488

Calculate the commutation M_{28}.

12. Given the following functions valued at $i = 0.03$:

x	$\ddot{a}_x$
72	8.06
73	7.73
74	7.43
75	7.15

Calculate p_{73}.

13. Given the following information for a 3-year life annuity due, contingent on the life of (x):

t	Payment	p_{x+t}
0	2	0.80
1	3	0.75
2	4	0.50

Assume that $i = 0.10$. Calculate the probability that the present value of the indicated payments exceeds 4.

14. Given $l_x = 100,000(100 - x)$, $0 \leq x \leq 100$ and $i = 0$. Calculate the present value of a whole life annuity issued to (80). The annuity is paid continuously at an annual rate of 1 per year the first year and 2 per year thereafter.

15. As in exercise 14, $l_x = 100,000(100 - x)$, $0 \leq x \leq 100$ and $i = 0$. Calculate the present value of a temporary 5-year life annuity issued to (80). The annuity is paid continuously at an annual rate of 1 per year the first year and 2 per year for four years thereafter.

16. Given $\delta = 0$, $\displaystyle\int_0^\infty t \, {}_tp_x dt = g$, and $\text{Var}(\bar{a}_{\overline{T}|}) = h$, where T is the future lifetime random variable for (x). Express $E[T]$ in terms of g and h.

17. Given:

x	69	70	71	72	$\cdots$	79	80	81	82
S_x	77,938	67,117	57,520	49,043	$\cdots$	13,483	10,875	8,691	6,875

Calculate $(D\ddot{a})_{70:\overline{10}|}$ which denotes the present value of a decreasing annuity. The first payment of 10 is at age 70, the second of 9 is scheduled for age 71, and so on. The last payment of 1 is scheduled for age 79.

18. Show that

$$\frac{a_{\overline{1}|}S_x - \ddot{a}_{\overline{2}|}S_{x+1} + \ddot{a}_{\overline{1}|}S_{x+2}}{D_x}$$

simplifies to A_x.

19. For a force of interest of $\delta > 0$, the value of $E\left(\bar{a}_{\overline{T}|}\right)$ is equal to 10. With the same mortality, but a force of interest of 2δ, the value of $E\left(\bar{a}_{\overline{T}|}\right)$ is 7.375. Also $\text{Var}(\bar{a}_{\overline{T}|}) = 50$. Calculate $\bar{A}_x$.

20. Calculate $\bar{a}_{x+u}$ using the Illustrative Life Table at 5% for age $x + u = 35.75$. *Assumption a* applies.

C.4.2   Spreadsheet Exercises

1. Calculate $\ddot{a}_x$ based on the Illustrative Life Table at $i = 5\%$. Use the recursion formula (4.6.1). Construct a graph showing the values of $\ddot{a}_x$ for $i = 0, 2.5\%, 5\%, 7.5\%, 10\%$ and $x = 0, 1, 2, \ldots, 99$.

2. Consider again the structured settlement annuity mentioned in exercise 7 of Section C.1. In addition to the financial data and the scheduled payments, include now the information that the payments are contingent upon the survival of a life subject to the mortality described in exercise 3 of Section C.2. Calculate the sum of market values of bonds required to hedge the expected value of the annuity payments.

3. A life age $x = 50$ is subject to a force of mortality ν_{50+t} obtained from the force of mortality standard as follows:

$$\nu_{50+t} = \begin{cases} \mu_{50+t} + c & \text{for } 0 \leq t \leq 15 \\ \mu_{50+t} & \text{otherwise} \end{cases}$$

where μ_{50+t} denotes the force of mortality underlying the Illustrative Life Table. The force of interest is constant $\delta = 4\%$. Calculate the variance of the present value of an annuity immediate of one per annum issued to (50) for values of $c = -0.01, -0.005, 0, 0.005$, and 0.01. Draw the graph.

4. Create a spreadsheet which calculates $\ddot{a}_{x+u}^{(m)}$ and A_{x+u} for a given age, $x+u$, with x an integer and $0 \leq u \leq 1$, and a given interest rate i. Assume that mortality follows the Illustrative Life Table. Use formulas (4.8.5) and (4.3.5) (or (4.8.6) and (4.3.5) if you like.) for the annuity and analogous ones for the life insurance.

5. Use your spreadsheet's built-in random number feature to simulate 200 values of $Y = 1 + v + \cdots + v^K = \ddot{a}_{\overline{K+1}}$ where $K = K(40)$. Use $i = 5\%$ and assume mortality follows the Illustrative Life Table. Compare the sample mean and variance to the values given by formulas (4.2.7) and (4.2.9).

C.5   Net Premiums

C.5.1   Notes

The exercises sometimes use the notation based on the system of International Actuarial Notation. Appendix 4 of *Actuarial Mathematics* by Bowers *et al.* describes the system. Here are the premium symbols and definitions used in these exercises.

$\overline{P}\left(\overline{A}_x\right)$ denotes the annual rate of payment of net premium, paid continuously, for a whole life insurance of 1 issued on the life of (x), benefit paid at the moment of death.

$\overline{P}\left(\overline{A}_{x:\overline{n}|}\right)$ denotes the annual rate of payment of net premium for an endowment insurance of 1 issued on the life of (x). The death benefit is paid at the moment of death.

A life insurance policy is fully continuous if the death benefit is paid at the moment of death, and the premiums are paid continuously over the premium payment period.

Policies with limited premium payment periods can be described symbolically with a pre-subscript. For example, $_nP\left(\overline{A}_x\right)$ denotes the annual rate of payment of premium, paid once per year, for a whole life insurance of 1 issued on the life of (x), benefit paid at the moment of death. For a policy with the death benefit paid at the end of the year of death the symbol is simplified to $_nP_x$.

C.5.2   Theory Exercises

1. Given: $_{20}P_{25} = 0.046$, $P_{25:\overline{20}|} = 0.064$, and $A_{45} = 0.640$. Calculate $P^1_{25:\overline{20}|}$.

2. A level premium whole life insurance of 1, payable at the end of the year of death, is issued to (x). A premium of G is due at the beginning of each year, provided (x) survives. Given:

(i) L = the insurer's loss when $G = P_x$

(ii) L^* = the insurer's loss when G is chosen such that $E[L^*] = -0.20$

(iii) $\mathrm{Var}[L] = 0.30$

Calculate $\mathrm{Var}[L^*]$.

3. Use the Illustrative Life Table and $i = 5\%$ to calculate the level net annual premium payable for ten years for a whole life insurance issued to a person age 25. The death benefit is 50,000 initially, and increases by 5,000 at ages $30, 35, 40, 45$ and 50 to an ultimate value of 75,000. Premiums are paid at the beginning of the year and the death benefits are paid at the end of the year.

4. Given the following values calculated at $d = 0.08$ for two whole life policies issued to (x):

	Death Benefit	Premium	Variance of Loss
Policy A	4	0.18	3.25
Policy B	6	0.22	

Premiums are paid at the beginning of the year and the death benefit sare paid at the end of the year. Calculate the variance of the loss for policy B.

5. A whole life insurance issued to (x) provides 10,000 of insurance. Annual premiums are paid at the beginning of the year for 20 years. Death claims are paid at the end of the year of death. A premium refund feature is in effect during the premium payment period which provides that one half of the last premium paid to the company is refunded as an additional death benefit. Show that the net annual premium is equal to

$$\frac{10,000A_x}{(1+d/2)\ddot{a}_{x:\overline{20}|} - \left(1 - v^{20}{}_{20}p_x\right)/2}.$$

6. Obtain an expression for the annual premium $_nP_x$ in terms of net single premiums and the rate of discount d. ($_nP_x$ denotes the net annual premium payable for n years for a whole life insurance issued to x.)

7. A whole life insurance issued to (x) provides a death benefit in year j of $b_j = 1,000(1.06)^j$ payable at the end of the year. Level annual premiums are payable for life. Given: $1,000P_x = 10$ and $i = 0.06$ per year. Calculate the net annual premium.

8. Given:

(i) $A_x = 0.25$

(ii) $A_{x+20} = 0.40$

(iii) $A_{x:\overline{20}|} = 0.55$

(iv) $i = 0.03$

(v) *assumption a* applies.

Calculate $1000P\left(\bar{A}_{x:\overline{20}|}\right)$.

9. A fully continuous whole life insurance of 1 is issued to (x). Given:

(i) The insurer's loss random variable is $L = v^T - \bar{P}\left(\bar{A}_x\right)\bar{a}_{\overline{T}|}$.

(ii) The force of interest δ is constant.

(iii) The force of mortality is constant: $\mu_{x+t} = \mu, t \geq 0$.

Show that $\text{Var}(L) = \mu/(2\delta + \mu)$.

10. A fully-continuous level premium 10-year term insurance issued to (x) pays a benefit at death of 1 plus the return of all premiums paid accumulated with interest. The interest rate used in calculating the death benefit is the same as that used to determine the present value of the insurer's loss. Let G denote the rate of annual premium paid continuously.

(a) Write an expression for the insurer's loss random variable L.

(b) Derive an expression for $\text{Var}[L]$.

(c) Show that, if G is determined by the equivalence principle, then

$$\text{Var}[L] = {}^{2}\bar{A}^{1}_{x:\overline{10}|} + \frac{\left(\bar{A}^{1}_{x:\overline{10}|}\right)^{2}}{{}_{10}p_{x}}.$$

The pre-superscript indicates that the symbol is evaluated at a force of interest of 2δ, where δ is the force of interest underlying the usual symbols.

11. Given:

(i) $i = 0.10$

(ii) $a_{30:\overline{9}|} = 5.6$

(iii) $v^{10}\,{}_{10}p_{30} = 0.35$

Calculate $1000\,P^{1}_{30:\overline{10}|}$

12. Given:

(i) $i = 0.05$

(ii) $10,000A_{x} = 2,000.$

Apply *assumption a* and calculate $10,000\bar{P}\left(\bar{A}_{x}\right) - 10,000P\left(\bar{A}_{x}\right)$.

13. Show that

$$1 - \frac{\left(P_{30:\overline{15}|} - {}_{15}P_{30}\right)\ddot{a}_{30:\overline{15}|}}{v^{15}\,{}_{15}p_{30}}$$

simplifies to A_{45}.

14. Given:

(i) $\bar{A}_{x} = 0.3$

(ii) $\delta = 0.07.$

A whole life policy issued to (x) has a death benefit of 1,000 paid at the moment of death. Premiums are paid twice per year. Calculate the semi-annual net premium using *assumption a*.

15. Given the following information about a fully continuous whole life insurance policy with death benefit 1 issued to (x):

(i) The net single premium is $\bar{A}_{x} = 0.4$.

(ii) $\delta = 0.06$

(iii) $\text{Var}[L] = 0.25$ where L denotes the insurer's loss associated with the net annual premium $\bar{P}\left(\bar{A}_{x}\right)$.

Under the same conditions, except that the insurer requires a premium rate of $G = 0.05$ per year paid continuously, the insurer's loss random variable is L^*. Calculate $\text{Var}[L^*]$.

16. A fully discrete 20-year endowment insurance of 1 is issued to (40). The insurance also provides for the refund of all net premiums paid accumulated at the interest rate i if death occurs within 10 years of issue. Present values are calculated at the same interest rate i. Using the equivalence principle, the net annual premium payable for 20 years for this policy can be written in the form:

$$\frac{A_{40:\overline{20}|}}{k}$$

Determine k.

17. L is the loss random variable for a fully discrete, 2-year term insurance of 1 issued to (x). The net level annual premium is calculated using the equivalence principle. Given:

(i) $q_x = 0.1$,

(ii) $q_{x+1} = 0.2$ and

(iii) $v = 0.9$.

Calculate $\text{Var}(L)$.

18. Given:

(i) $\bar{A}^1_{x:\overline{n}|} = 0.4275$

(ii) $\delta = 0.055$, and

(iii) $\mu_{x+t} = 0.045, t \geq 0$

Calculate $1,000 \bar{P}\left(\bar{A}_{x:\overline{n}|}\right)$.

19. A 4-year automobile loan issued to (25) is to be repaid with equal annual payments at the end of each year. A four-year term insurance has a death benefit which will pay off the loan at the end of the year of death, including the payment then due. Given:

(i) $i = 0.06$ for both the actuarial calculations and the loan,

(ii) $\ddot{a}_{25:\overline{4}|} = 3.667$, and

(iii) $_4q_{25} = 0.005$.

(a) Express the insurer's loss random variable in terms of K, the curtate future lifetime of (25), for a loan of 1,000 assuming that the insurance is purchased with a single premium of G.

(b) Calculate G, the net single premium rate per 1,000 of loan value for this insurance.

(c) The automobile loan is 10,000. The buyer borrows an additional amount to pay for the term insurance. Calculate the total annual payment for the loan.

20. A level premium whole life insurance issued to (x) pays a benefit of 1 at the end of the year of death. Given:

(i) $A_x = 0.19$

(ii) $^2A_x = 0.064$, and

(iii) $d = 0.057$

Let G denote the rate of annual premium to be paid at the beginning of each year while (x) is alive.
(a) Write an expression for the insurer's loss random variable L.
(b) Calculate $E[L]$ and $Var[L]$, assuming $G = 0.019$.
(c) Assume that the insurer issues n independent policies, each having $G = 0.019$. Determine the minimum value of n for which the probability of a loss on the entire portfolio of policies is less than or equal to 5%. Use the normal approximation.

C.5.3   Spreadsheet Exercises

1. Reproduce the Illustrative Life Table values of C_x. Calculate M_x recursively, from the end of the table where $M_{99} = C_{99}$, using the relation $M_x = C_x + M_{x-1}$. Calculate the values of $S_x = M_x + M_{x+1} + \ldots$ using the same technique.

2. Use the Illustrative Life Table to calculate the initial net annual premium for a whole life insurance policy issued at age $x = 30$. The benefit is inflation protected: each year the death benefit and the annual premium increase by a factor of $1 + j$, where $j = 0.06$. Calculate the initial premium for interest rates of $i = 0.05, 0.06, 0.07$ and 0.08. Draw the graph of the initial premium as a function of i .

3. Use $i = 4\%$, the Illustrative Life Table, and the utility function $u(x) = (1 - e^{-ax})/a, a = 10^{-6}$, to calculate annual premiums for 10-year term insurance, issue age 40, using formula (5.2.9) : $E[u(-L)] = u(0)$. Display your results in a table with the sum insured C, the calculated premium, and the ratio to the net premium (loading). Draw the graph of the loading as a function of the sum insured. Do the same for premiums based on $a = 10^{-4}, 10^{-5}, 10^{-7}$ and 10^{-8} also. Show all the graphs on the same chart.

4. A whole life policy is issued at age 10 with premiums payable for life. If death occurs before age 15, the death benefit is the return of net premiums paid with interest to the end of the year of death. If death occurs after age 15, the death benefit is 1000. Calculate the net annual premium. Use the Illustrative Life Table and $i = 5\%$. Convince yourself that the net premium is independent of q_x for $x < 15$. (This problem is based on problem 21 at the end of Chapter 4 of *Life Contingencies* by C. W. Jordan.)

5. A 20-year term insurance is issued at age 45 with a face amount of 100,000. The net premium is determined using $i = 5\%$ and the Illustrative Life Table. The benefit is paid at the end of the year. Net premiums are invested in a fund earning j per annum and returned at age 65 if the insured survives. Calculate the net premium for values of j running from 5% to 9% in increments of 1%.

6. Determine the percentage z of annual salary a person must save each year in order to provide a retirement income which replaces 50% of final salary. Assume that the person is age 30, that savings earn 5% per annum, that salary increases at a rate of $j = 6\%$ per year, and that mortality follows the Illustrative Life Table. Draw the graph of z as a function of j running from 3% to 7% in increments of 0.5%

7. Mortality historically has improved with time. Let q_x denote the mortality table when a policy is issued. Suppose that the improvement (decreasing q_x) is described by $k^t q_x$ where t is the number of years since the policy was issued and k is a constant, $0 < k < 1$. Calculate the ratio of net premiums on the initial mortality basis to net premiums adjusted for $t = 10$ years of mortality improvement. Use $x = 30, k = 0.99$, the Illustrative Life Table for the initial mortality and $i = 5\%$.

C.6   Net Premium Reserves

Here are the additional symbols and definitions for reserves used in these exercises. Policies with premiums paid at the beginning of the year and death benefits paid at the end of the year of death are called fully discrete policies. Policies with premiums paid continuously and death benefits paid at the moment of death are called fully continuous.

$_k\overline{V}\left(\bar{A}_x\right)$ denotes the net premium reserve at the end of year k for a fully continuous whole life policy issued to (x).

$_k\overline{V}\left(\bar{A}_{x:\overline{n}|}\right)$ denotes the net premium reserve at the end of year k for a fully continuous n-year endowment insurance policy issued to (x).

Policies with limited premium payment periods can be described symbolically with a pre-superscript. For example, $_k^n V(\bar{A}_x)$ denotes the net premium reserve at the end of year k for an n-payment whole life policy issued to (x) with the benefit of 1 paid at the moment of death. Note that the corresponding net premium is denoted $_nP(\bar{A}_x)$.

C.6.1   Theory Exercises

1. A 20-year fully discrete endowment policy of 1000 is issued at age 35 on the basis of the Illustrative Life Table and $i = 5\%$. Calculate the amount of reduced paid-up insurance available at the end of year 5, just before the sixth premium is due. Assume that the entire reserve is available to fund the paid-up policy as described in section 6.8.

2. Given: $_{10}V_{25} = 0.1$ and $_{10}V_{35} = 0.2$. Calculate $_{20}V_{25}$.

3. Given: $_{20}V(\bar{A}_{40}) = 0.3847$, $\bar{a}_{40} = 20.00$, and $\bar{a}_{60} = 12.25$. Calculate $_{20}\overline{V}\left(\bar{A}_{40}\right) - _{20}V\left(\bar{A}_{40}\right)$.

4. Given the following information for a fully discrete 3-year special endowment insurance issued to (x):

k	c_{k+1}	q_{x+k}
0	2	0.20
1	3	0.25
2	4	0.50

Level annual net premiums of 1 are paid at the beginning of each year while (x) is alive. The special endowment amount is equal to the net premium reserve for year 3. The effective annual interest rate is $i = 1/9$. Calculate the end of policy year reserves recursively using formula (6.3.4) from year one with $_0V = 0$.

5. Given: $i = 0.06, q_x = 0.65, q_{x+1} = 0.85$, and $q_{x+2} = 1.00$. Calculate $_1V_x$. (Hint due to George Carr 1989: Calculate the annuity values recursively from $\ddot{a}_{x+2}$ back to $\ddot{a}_x$. Use (6.5.3).)

6. A whole life policy for 1000 is issued on May 1, 1978 to (60). Given:

(i) $i = 6\%$

(ii) $q_{70} = 0.033$

(iii) $1000\,_{10}V_{60} = 231.14$

(iv) $1000P_{60} = 33.00$, and

(v) $1000\,_{11}V_{60} = 255.40$

A simple method widely used in practice is used to approximate the reserve on December 31, 1988. Calculate the approximate value.

7. Given: For $k = 0, 1, 2, \ldots$ $_{k|}q_x = (0.5)^{k+1}$. Show that the variance of the loss random variable L for a fully discrete whole life insurance for (x) is

$$\frac{1}{2}\left(\frac{v^2}{2 - v^2}\right)$$

where $v = (1 + i)^{-1}$.

8. Given, for a fully discrete 20-year deferred life annuity of 1 per year issued to (35):

(i) Mortality follows the Illustrative Life Table.

(ii) $i = 0.05$

(iii) Level annual net premiums are payable for 20 years.

Calculate the net premium reserve at the end of 10 years for this annuity.

9. A special fully discrete 2-year endowment insurance with a maturity value of 1000 is issued to (x). The death benefit in each year is 1000 plus the net level premium reserve at the end of that year. Given $i = 0.10$ and the following data:

k	q_{x+k}	c_{k+1}	$_kV$
0	0.10	$1000 + {}_1V$	0
1	0.11	2000	?
2			1000

Calculate the net level premium reserve $_1V$.

10. Use the Illustrative Life Tables and $i = 0.05$ to calculate $1000\,_{15}V_{45:\overline{20}|}$.

11. Use the Illustrative Life Tables and $i = 0.05$ to calculate $1000\,_{15}V^1_{45:\overline{20}|}$.

12. Given the data in exercise 4, calculate the variance of the loss Λ_1 allocated to policy year two.

13. Given:

k	$\ddot{a}_{\overline{k}|}$	$_{k-1|}q_x$
1	1.000	0.33
2	1.930	0.24
3	2.795	0.16
4	3.600	0.11

Calculate $_2V_{x:\overline{4}|}$.

14. A fully discrete whole life policy with a death benefit of 1000 is issued to (40). Use the Illustrative Life Table and $i = 0.05$ to calculate the variance of the loss allocated to policy year 10.

15. At an interest rate of $i = 4\%$, $_{23}^{20}V_{15} = 0.585$ and $_{24}^{20}V_{15} = 0.600$. Calculate p_{38}.

16. A fully discrete whole life insurance is issued to (x). Given: $P_x = 4/11$, $_tV_x = 0.5$, and $\ddot{a}_{x+t} = 1.1$. Calculate i.

17. For a special fully discrete whole life insurance of 1000 issued on the life of (75), increasing net premiums, Π_k, are payable at time k, for $k = 0, 1, 2, \ldots$. Given:

(i) $\Pi_k = \Pi_0(1.05)^k$ for $k = 0, 1, 2, \ldots$

(ii) Mortality follows de Moivre's law with $\omega = 105$.

(iii) $i = 5\%$

Calculate the net premium reserve at the end of policy year five.

18. Given for a fully discrete whole life insurance for 1500 with level annual premiums on the life of (x):

(i) $i = 0.05$

(ii) The reserve at the end of policy year h is 205.

(iii) The reserve at the end of policy year $h - 1$ is 179.

(iv) $\ddot{a}_x = 16.2$

Calculate $1000q_{x+h-1}$

19. Given:

(i) $1 + i = (1.03)^2$

(ii) $q_{x+10} = 0.08$

(iii) $1000_{10}V_x = 311.00$

(iv) $1000P_x = 60.00$

(v) $1000_{11}V_x = 340.86$

(a) Approximate $1000_{10.5}V_x$ by use of the traditional rule: interpolate between reserves at integral durations and add the unearned premium.
(b) *Assumption a* applies. Calculate the exact value of $1000_{10.5}V_x$.

20. Given: $q_{31} = 0.002$, $\ddot{a}_{32:\overline{13}|} = 9$, and $i = 0.05$. Calculate $_1V_{31:\overline{14}|}$.

C.6.2 Spreadsheet Exercises

1. Calculate a table of values of $_tV_{30}$ for $t = 0, 1, 2, \ldots, 69$, using the Illustrative Life Table and $i = 4\%$. Recalculate for $i = 6\%$ and 8%. Draw the three graphs of $_tV_{30}$ as a function of t, corresponding to $i = 4\%, 6\%$, and 8%. Put the graphs on a single chart.

2. A 10-year endowment insurance with a face amount of 1000 is issued to (50). Calculate the savings Π_k^r and risk Π_k^s components of the net annual premium $1000P_{50:\overline{10}|}$ (formulas 6.3.6 and 6.3.7) over the life of the policy. Use the Illustrative Life Table and $i = 4\%$. Draw the graph of Π_k^s as a function of the policy year k. Investigate its sensitivity to changes in i by calculating the graphs for $i = 1\%$ and 7% and showing all three on a single chart.

3. A 10,000 whole life policy is issued to (30) on the basis of the Illustrative Life Table at 5%. The actual interest earned in policy years 1 - 5 is $i' = 6\%$. Assume the policyholder is alive at age 35 and the policy is in force.
(a) Calculate the technical gain realized in each year using method 2 (page 69).
(b) Calculate the accumulated value of the gains (using $i' = 6\%$) at age 35.
(c) Determine the value of i' (level over five years) for which the accumulated gains are equal to 400.

4. This exercise concerns a flexible life policy as described in section 6.8. The policyholder chooses the benefit level c_{k+1} and the annual premium Π_k at the beginning of each policy year $k+1$. The choices are subject to these constraints:

(i) $\Pi_0 = 100,000P_x$, the net level annual premium for whole life in the amount of 100,000.

(ii) $0 \leq \Pi_{k+1} \leq 1.2\Pi_k$ for $k = 0, 1, \ldots$

(iii) $c_1 \leq c_{k+1} \leq 1.2c_k$ for $k = 1, 2, \ldots$

(iv) $_{k+1}V \geq 0$ for $k = 0, 1, 2, \ldots$

The initial policyholder's account value is $_0V = 0$. Thereafter the policyholder's account values accumulate according to the recursion relation (6.3.4) with the interest rate specified in the policy as $i = 5\%$ and mortality following the Illustrative Life Table with $x = 40$. Investigate the insurer's cumulative gain on the policy under two scenarios:

(S_1) The policyholder attempts to maximize insurance coverage at minimal costs over the first five policy years. The strategy is implemented by choosing $c_{k+1} = 1.2c_k$ for $k = 1, 2, \ldots$ and choosing the level premium rate which meets the constraints but has $_5V = 0$. Calculate the insurer's annual gains assuming $i' = 5.5\%$ and the policyholder dies during year 5.

(S_2) The policyholder elects to maximize savings by choosing minimum coverage and maximum premiums. Calculate the present value of the insurer's annual gains assuming $i' = 5.5\%$ and the policyholder survives to the end of year 5.

C.7   Multiple Decrements: Exercises

C.7.1   Theory Exercises

1. In a double decrement table $\mu_{1,x+t} = 0.01$ for all $t \geq 0$ and $\mu_{2,x+t} = 0.02$ for all $t \geq 0$. Calculate $q_{1,x}$.

2. Given $\mu_{j,x+t} = j/150$ for $j = 1, 2, 3$ and $t > 0$. Calculate $E[T \mid J = 3]$.

3. A whole life insurance policy provides that upon accidental death as a passenger on public transportation a benefit of 3000 will be paid. If death occurs from other accidental causes, a death benefit of 2000 will be paid. Death from causes other than accidents carries a benefit of 1000. Given, for all $t \geq 0$:

(i) $\mu_{j,x+t} = 0.01$ where $j = 1$ indicates accidental death as a passenger on public transportation.

(ii) $\mu_{j,x+t} = 0.03$ where $j = 2$ indicates accidental death other than as a passenger on public transportation.

(iii) $\mu_{j,x+t} = 0.03$ where $j = 3$ indicates non-accidental death.

(iv) $\delta = 0.03$.

Calculate the net annual premium for issue age x assuming continuous premiums and immediate payment of claims.

4. In a double decrement table, $\mu_{1,x+t} = 1$ and $\mu_{2,x+t} = \frac{t}{t+1}$ for all $t \geq 0$. Calculate

$$m_x = \frac{q_x}{\int_0^1 {}_t p_x \, dt}.$$

5. A two year term policy on (x) provides a benefit of 2 if death occurs by accidental means and 1 if death occurs by other means. Benefits are paid at the moment of death. Given for all $t \geq 0$:

(i) $\mu_{1,x+t} = t/20$ where 1 indicates accidental death.

(ii) $\mu_{2,x+t} = t/10$ where 2 indicates other than accidental death.

(iii) $\delta = 0$

Calculate the net single premium.

6. A multiple decrement model has 3 causes of decrement. Each of the decrements has a uniform distribution over each year of age so that the equation (7.3.4) holds for at all ages and durations. Given:

(i) $\mu_{1,30+0.2} = 0.20$

(ii) $\mu_{2,30+0.4} = 0.10$

(iii) $\mu_{3,30+0.8} = 0.15$

Calculate q_{30}.

7. Given for a double decrement table:

x	$q_{1,x}$	$q_{2,x}$	p_x
25	0.01	0.15	0.84
26	0.02	0.10	0.88

(a) For a group of 10,000 lives aged $x = 25$, calculate the expected number of lives who survive one year and fail due to decrement $j = 1$ in the following year.
(b) Calculate the effect on the answer for (a) if $q_{2,25}$ changes from 0.15 to 0.25.

8. Given the following data from a double decrement table:

(i) $q_{1,63} = 0.050$

(ii) $q_{2,63} = 0.500$

(iii) $_{1|}q_{63} = 0.070$

(iv) $_{2|}q_{1,63} = 0.042$

(v) $_3p_{63} = 0$.

For a group of 500 lives aged $x = 63$, calculate the expected number of lives who will fail due to decrement $j = 2$ between ages 65 and 66.

9. Given the following for a double decrement table:

(i) $\mu_{1,x+0.5} = 0.02$

(ii) $q_{2,x} = 0.01$

(iii) Each decrement is uniformly distributed over each year of age, thus (7.3.4) holds for each decrement.

Calculate $1000q_{1,x}$.

10. A multiple decrement table has two causes of decrement: (1) accident and (2) other than accident. A fully continuous whole life insurance issued to (x) pays c_1 if death results by accident and c_2 if death results other than by accident. The force of decrement 1 is a positive constant μ_1. Show that the net annual premium for this insurance is $c_2\bar{P}_x + (c_1 - c_2)\mu_1$.

C.8   Multiple Life Insurance: Exercises

C.8.1   Theory Exercises

1. The following excerpt from a mortality table applies to each of two independent lives (80) and (81):

x	q_x
80	0.50
81	0.75
82	1.00

Assumption a applies. Calculate $q_{80:81}^1$, $q_{80:\overset{2}{81}}$, $q_{80:81}$ and $q_{\overline{80:81}}$.

2. Given:

(i) $\delta = 0.055$

(ii) $\mu_{x+t} = 0.045, t \geq 0$

(iii) $\mu_{y+t} = 0.035, t \geq 0$

Calculate $\bar{A}_{xy}^2$ as defined by formula (8.8.8).

3. In a certain population, smokers have a force of mortality twice that of non-
smokers. For non-smokers, $s(x) = 1 - x/75$, $0 \leq x \leq 75$. Calculate $\overset{\circ}{e}_{55:65}$ for a
smoker (55) and a non-smoker (65) .

4. A fully-continuous insurance policy is issued to (x) and (y). A death benefit
of 10,000 is payable upon the second death. The premium is payable continuously until the last death. The annual rate of payment of premium is c while
(x) is alive and reduces to $0.5c$ upon the death of (x) if (x) dies before (y). The
equivalence principle is used to determine c. Given:

(i) $\delta = 0.05$

(ii) $\bar{a}_x = 12$

(iii) $\bar{a}_y = 15$

(iv) $\bar{a}_{xy} = 10$

Calculate c.

5. A fully discrete last-survivor insurance of 1 is issued on two independent lives
each age x. Level net annual premiums are paid until the first death. Given:

(i) $A_x = 0.4$

(ii) $A_{xx} = 0.55$

(iii) $a_x = 9.0$

Calculate the net annual premium.

6. A whole life insurance pays a death benefit of 1 upon the second death of (x) and (y). In addition, if (x) dies before (y), a payment of 0.5 is payable at the time of death. Mortality for each life follows the Gompertz law with a force of mortality given by $\mu_z = Bc^z$, $z \geq 0$. Show that the net single premium for this insurance is equal to

$$\bar{A}_x + \bar{A}_y - \bar{A}_w \left(1 - 0.5c^{x-w}\right)$$

where $c^w = c^x + c^y$.

7. Given:

(i) Male mortality has a constant force of mortality $\mu = 0.04$.

(ii) Female mortality follows de Moivre's law with $\omega = 100$.

Calculate the probability that a male age 50 dies after a female age 50.

8. Given:

(i) Z is the present-value random variable for an insurance on the independent lives of (x) and (y) where

$$Z = \begin{cases} v^{T(y)} & \text{if } T(y) > T(x) \\ 0 & \text{otherwise} \end{cases}$$

(ii) (x) is subject to a constant force of mortality of 0.07.

(iii) (y) is subject to a constant force of mortality of 0.09.

(iv) The force of interest is a constant $\delta = 0.06$.

Calculate Var$[Z]$.

9. A fully discrete last-survivor insurance of 1000 is issued on two independent lives each age 25. Net annual premiums are reduced by 40% after the first death. Use the Illustrative Life Table and $i = 0.05$ to calculate the initial net annual premium.

10. A life insurance on John and Paul pays death benefits at the end of the year of death as follows:

(i) 1 at the death of John if Paul is alive,

(ii) 2 at the death of Paul if John is alive,

(iii) 3 at the death of John if Paul is dead and

(iv) 4 at the death of Paul if John is dead.

The joint distribution of the lifetimes of John and Paul is equivalent to the joint distribution of two independent lifetimes each age x. Show that the net single premium of this life insurance is equal to $7\bar{A}_x - 2\bar{A}_{xx}$.

C.8.2   Spreadsheet Exercises

1. Use the Illustrative Life Table and $i = 5\%$ to calculate the joint life annuity, $a_{x:y}$, the joint-and-survivor annuity, $a_{\overline{x:y}}$, and the reversionary annuity, $a_{x/y}$, for independent lives lives age $x = 65$ and $y = 60$.

2. (8.4.8) A joint-and-survivor annuity is payable at the rate of 10 per year at the end of each year while either of two independent lives (60) and (50) is alive. Given:

(i) The Illustrative Life Table applies to each life.

(ii) $i = 0.05$

Calculate a table of survival probabilities for the joint-and-survivor status. Use it to calculate the variance of the annuity's present value random variable.

3. Use the Illustrative Life Table and $i = 5\%$ to calculate the net level annual premium for a second-to-die life insurance on two independent lives age (35) and (40). Assume that premiums are paid at the beginning of the year as long as both insured lives survive. The death benefit is paid at the end of the year of the second death.

4. Calculate the net premium reserve at the end of years 1 through 10 for the policy in exercise 3. Assume that the younger life survives 10 years and that the older life dies in the sixth policy year.

5. Given:

(i) $\mu_x = A + Bc^x$ for $x \geq 0$ where $A = 0.004$, $B = 0.0001$, $c = 1.15$, and

(ii) $\delta = 5\%$.

Approximate $\bar{a}_{30:40}$ and $\bar{A}^1_{30:40}$.

C.9 The Total Claim Amount in a Portfolio

C.9.1 Theory Exercises

1. The claim made in respect of policy h is denoted S_h. The three possible values of S_h are as follows:

$$S_h = \begin{cases} 0 & \text{if the insured life } (x) \text{ survives,} \\ 100 & \text{if the insured surrenders the policy, and} \\ 1000 & \text{if the insured dies.} \end{cases}$$

The probability of death is $q_{1,x} = 0.001$, the probability of surrender is $q_{2,x} = 0.15$, and the probability of survival is $p_x = 1 - q_{1,x} - q_{2,x}$. Use the normal approximation to calculate the probability that the aggregate claims of five identically distributed policies $S = S_1 + \cdots + S_5$ exceeds 200.

2. The aggregate claims S are approximately normally distributed with mean μ and variance σ^2. Show that the stop-loss reinsurance net premium $\rho(\beta) = E[(X - \beta)^+]$ is given by

$$\rho(\beta) = (\mu - \beta)\Phi\left(\frac{\mu - \beta}{\sigma}\right) + \sigma\phi\left(\frac{\mu - \beta}{\sigma}\right)$$

where Φ and ϕ are the standard normal distribution and density functions.

3. Consider the compound model described by formula (9.4.6): $S = X + .. + X_N$ where N, X_i are independent, and X_i are identically distributed. Show that the moment generating function of S is $M_S(t) = M_N(\log(M_X(t)))$ where $M_N(t)$ and $M_X(t)$ are the moment generating functions of N and X. This provides a means of estimating moments of S from estimates of moments of X and N. For example, $E[S] = E[N]E[X]$ and

$$E[S^2] = E[N^2]E[X]^2 + E[N]\left(E[X^2] - E[X]^2\right).$$

4. A reinsurance contract provides a payment of

$$R = \begin{cases} S - \beta & \text{if } \beta < S < \gamma \\ \gamma - \beta & \text{if } S \geq \gamma \end{cases}$$

Express $E[R]$ in terms of the cummulative distribution function of S.

5. (a)   Express $F(x)$ in terms of the function $\rho(\beta)$.

   (b)   Given that $\rho(\beta) = \left(2 + \beta + \frac{1}{4}\beta^2\right), \beta \geq 0$, find $F(x)$ and $f(x)$.

6. Suppose that $f(0), f(1), f(2), \ldots$ are probabilities for which the following holds:

$$f(1) = 3f(0), f(2) = 2f(0) + 1.5f(1),$$
$$f(x) = \frac{1}{x}(3f(x-3) + 4f(x-2) + 3f(x-1)) \text{ for } x = 3, 4, \ldots$$

What is the value of $f(0)$?

7. Suppose that $\log S$ is normally distributed with parameters, μ and σ. Calculate the net stop-loss premium $\rho(\beta) = E[(S - \beta)^+]$ for a deductible β.

8. (a)   For the portfolio defined by (9.3.5), calculate the distribution of aggregate claims by applying the method of dispersion with a span of 0.5.

(b)   Apply the compound Poisson approximation with the same discetization.

C.10 Expense Loadings

C.10.1 Theory Exercises

1. Consider the endowment policy of section 6.2, restated here for convenience: sum insured $= 1000$, duration $n = 10$, initial age $x = 40$, De Moivre mortality with $\omega = 100$, and $i = 4\%$.

(i) The acquisition expense is 50. No other expenses are recognized ($\beta = \gamma = 0$). Calculate the expense-loaded annual premium and the expense-loaded premium reserves for each policy year.

(ii) Determine the maximum value of acquisition expense if negative expense-loaded reserves are to be avoided.

2. Give a verbal interpretation of $-_kV^\alpha$.

3. Consider the term insurance policy of section 6.2, restated here for convenience: sum insured $= 1000$, duration $n = 10$, initial age $x = 40$, De Moivre mortality with $\omega = 100$, and $i = 4\%$.

(i) The acquisition expense is 40. No other expenses are recognized ($\beta = \gamma = 0$). Calculate the expense-loaded annual premium and the expense-loaded premium reserves for each policy year.

(ii) If the expense-loaded premium reserves are not allowed to be negative, what is the insurer's initial investment for selling such a policy?

4. Calculate the components $1000P$, $1000P^\alpha$, $1000P^\beta$ and $1000P^\gamma$ of the expense-loaded premium $1000P^a$ for a whole life insurance of 1000 issued to a life age 35. The policy has level annual premiums for 30 years, becoming paid-up at age 65. The company has expenses as follows:

acquisition expense   12 at the time of issue,
collection expenses   15% of each **expense-loaded** premium, and
administration expens   1 at the beginning of each policy year.

Use the Illustrative Life Table and $i = 5\%$.

5. For the policy described in exercise 4, calculate components 1000_kV, 1000_kV^α, and 1000_kV^γ of the expense-loaded premium reserve 1000_kV^a for year $k = 10$.

C.10.2   Spreadsheet Exercises

1. Develop a spreadsheet to calculate the expense-loaded premium components and the expense-loaded premium reserve components for each policy year of a 20-year endowment insurance issued to a life age 40. Use the Illustrative Life Table and $i = 6\%$. Assume that acquisition expense is 20 per 1000 of insurance, collection expense is 5% of the expense-loaded premium, and administration expense is 3 at the beginning of each policy year.

C.11 Estimating Probabilities of Death

C.11.1 Theory Exercises

1. Consider the following two sets of data:

    (a) $D_x = 36$ $E_x = 4820$
    (b) $D_x = 360$ $E_x = 48200$

For each set, calculate a 90% confidence interval for q_x.

2. We model the uncertainty about θ (the unknown value of $\mu_{x+1/2}$) by a gamma distribution such that $E[\theta] = 0.007$ and $Var(\theta) = 0.000007$. An additional 36 deaths are observed for an additional exposure of 4820. Calculate our posterior expectation and standard deviation of θ, and our estimate for q_x.

3. Write down the equations from which

(i) λ^l and

(ii) λ^u are obtained.

(iii) Rewrite these equations in terms of integrals over $f(x; n)$, the probability density function of the gamma distribution with shape parameter n and scale parameter 1.

4. In a clinical experiment, a group of 50 rats is under observation until the 20th rat dies. At that time the group has lived a total of 27.3 rat years. Estimate the force of mortality (assumed to be constant) of this group of rats. What is their life expectancy?

5. A certain group of lives has a total exposure of 9758.4 years between ages x and $x + 1$. There were 357 deaths by cause one, 218 deaths by cause two, and 528 deaths by all other causes combined. Estimate the probability that a life age x will die by cause one within a year.

6. There are 100 life insurance policies in force, insuring lives age x. An additional 60 polices are issued at age $x + \frac{1}{4}$. Four deaths are observed between age x and $x + 1$; we assume that these deaths occur at age $x + 0.5$. Calculate the classical estimator given by formula (11.2.3), and the maximum likelihood estimator based on the *assumption b*, a constant force of mortality (11.4.2).

7. The force of mortality is constant over the year age $(x, x+1]$. Ten lives enter observation at age x. Two lives enter observation at age $x+0.4$. Two lives leave observation at age $x+0.8$, one leaves at age $x+0.2$ and one leaves at age $x+0.5$. There is one death at age $x + 0.6$. Calculate the maximum likelihood estimate of the force of mortality.

8. A double decrement model is used to study two causes of death in the interval of age $(x, x+1]$.

The forces of each cause are constant.

1,000 lives enter the study at age x.

40 deaths occur due to cause 1 in $(x, x+1]$.

50 deaths occur due to cause 2 in $(x, x+1]$.

Calculate the maximum likelihood estimators of the forces of decrement.

9. The Illustrative Life Table is used for a standard table in a mortality study. The study results in the following values of exposures E_x and deaths D_x over $[40, 45)$

x	E_x	D_x
40	1150	6
41	900	5
42	1200	12
43	1400	9
44	1300	13

Calculate the mortality ratio $\hat{f}$ and the 90% confidence interval for f. Calculate the estimates of $\hat{q}_{40}, \hat{q}_{41}, \ldots \hat{q}_{45}$ corresponding to $\hat{f}$.

Appendix D

Solutions

D.0   Introduction

We offer solutions to most theory exercises which we hope students will find useful. When the solution is straightforward we simply give the answer. For the spreadsheet exercises we describe the solution and give some values to use to verify your work. We leave the joy of writing the program to the student.

We have tried hard to avoid errors. We hope that students and other users who discover errors will inform us promptly. We are also interested in seeing elegant or insightful solutions and new problems.

The solutions occasionally refer to the Illustrated Life Table and its functions. They are in Appendix E.

D.1 Mathematics of Compound Interest

D.1.1 Solutions to Theory Exercises

1. This follows easily from equation (1.5.8).

2. Fix $i > 0$ and consider the function $f(x) = [(1+i)^x - 1]x^{-1} = (e^{\delta x} - 1)/x$. From the power series expansion

$$f(x) = \delta + \frac{1}{2}\delta^2 x + \frac{1}{3!}\delta^3 x^2 + \ldots,$$

it is easy to see that $f'(x) > 0$ for all $x > 0$. It follows that $f(x)$ increases from $f(0+) = \delta$ to $f(1) = i$. Therefore $g(x) = f(x^{-1})$ decreases on $[1, \infty)$ from i to δ. Thus, $i^{(m)} = g(m)$ decreases to δ as m increases. Similarly, $d^{(m)}$ increases to δ as m increases.

3. The accumulated value of the deposits as of January 1, 1999 is $Xs_{\overline{10}|0.06}$. The present value of the bond payments as of January 1,1999 is $15,000a_{\overline{5}|0.06}$. Equate the two values and solve for $X = 4794$.

4. Let i be the effective annual interest rate. Then $1 + i = (1 + j/2)^2$. The equation of value is

$$
\begin{aligned}
5.89 &= v^2 + v^4 + \ldots \\
     &= \frac{v^2}{1 - v^2}.
\end{aligned}
$$

Hence $(1 + j/2)^4 = 1 + 1/5.89$ and so $j = 8\%$.

5. Let $u = \frac{1+k}{1.04}$ and write the equation of value as follows:

$$
\begin{aligned}
51 &= \frac{1+k}{1.04} + \left(\frac{1+k}{1.04}\right)^2 + \ldots \\
   &= u + u^2 + \ldots \\
   &= \frac{u}{1 - u} \\
   &= \frac{1+k}{0.04 - k}.
\end{aligned}
$$

Solve for $k = 2\%$.

6. Use equation (1.9.8) with a starting value of $\delta = 12\%$. The price of the coupon payments is $p = 94 - 100(1.12)^{-10} = 61.80$. The sum of the payments is $r = 100$ and

$$a(\delta) = 5\frac{1 - e^{-10\delta}}{e^{\delta/2} - 1}.$$

The solution is $\delta = 9.94\%$. This is equivalent to 10.19\% per year convertible semiannually.

7. The equation of value is

$$
\begin{aligned}
1000 &= x(v + v^2 + v^3) + 3x(v^4 + v^5 + v^6) \\
&= x(3a_{\overline{6}|} - 2a_{\overline{3}|}) \\
&= x(11.504459)
\end{aligned}
$$

where the symbols correspond to a values of $i = 1\%$. So $x = 86.92$.

8. At the time of the loan,

$$
\begin{aligned}
4000 &= kv + 2kv^2 + 3kv^3 + \ldots + 30kv^{30} \\
&= k(Ia)_{\overline{30}|} \\
&= k\frac{\ddot{a}_{\overline{30}|} - 30v^{30}}{0.04}
\end{aligned}
$$

so $k = 18.32$. Note that the initial payment is less than the interest (160) required on the loan so the loan balance increases. Immediately after the ninth payment the outstanding payments, valued at the original loan interest rate, is found as follows:

$$
\begin{aligned}
10kv + 11kv^2 + \ldots + 30kv^{21} &= 9ka_{\overline{21}|} + k(Ia)_{\overline{21}|} \\
&= (18.32)(9(14.02916) + 134.37051) \\
&= 4774.80.
\end{aligned}
$$

9. Let $j = i^{(2)}/2$ and solve $98.51 = 2(1 + j)^{-1} + 102(1 + j)^{-2}$ for $(1 + j)^{-1} = 0.9729882$. This corresponds to $i^{(2)} = 5.55\%$.

10. From (i) and (ii) we get $12(120)\, a_{\overline{\infty}|}^{(12)} = 12(365.47)\, a_{\overline{n}|}^{(12)}$ and solve for $v^n = 0.6716557$. Now use (iii) and (iv) to solve for $X = 12000$.

D.1.2   Solutions to Spreadsheet Exercises

1. (a) The investment yield is 9.986%.

2. *Guide*: Set up a spreadsheet with a trial value of X. Since a total of $X + 2X + 3X + \cdots + 6X = 21X$ is withdrawn, a good trial value is about $100,000/20 = 5,000$. Use the fundamental formula (1.2.1) to calculate the fund balance at the end of each half-year. Then experiment with X until an end-of-period six balance of zero is found. $X = 6,128$.(Alternatively, in the last stage, use Goal Seek to determine the value of X which makes the target balance zero.) Note that the end-of-period six balance is the fund balance beginning the seventh half-year. Adapting notation of the text to half-years we have $F_0 = 100,000$, $F_1 = F_0\,(1.03)^2 - X$, $F_2 = F_1\,(1.03)^2 F_1 - 2X$, and so on.

3. *Guide*: Set up an amortization table using a spreadsheet and a trial value of $i = 0.03$. In a cell apart from the table, calculate the target $P - 6I$ for the fifth year. Then use Goal Seek to determine i so that the target cell is zero. $i = 10.93\%$.

4. *Guide*: The fund deposit X satisfies $Xs_{\overline{10}|:0.03} = 10,000$. In effect, the company accepts 10,000 now in exchange for 10 semiannual payments of $300+X$. Calculate X using the spreadsheet financial functions. The internal rate of return j equates the future cash flows $300+X$ to 10,000. Set up your spreadsheet with a trial value of j. Use the Goal Seek feature to detemine the value of j. $j = 7.80\%$

5. *Guide*: Put $i = 10\%$ and a trial value of X into cells. Calculate the net present value of the payments of 100 minus the payments of X in another cell as follows:

$$100\ddot{a}_{\overline{10}|} - Xv^{19}\ddot{a}_{\overline{\infty}|} = \frac{100 - v_{10}100 - v_{19}X}{d}$$

Use the Goal Seek feature to determine the valaue of X for which the resent value is zero. $X = 375.80$

6. *Guide*: Set up a spreadsheet to amortize the loan using a trial value of $X = 30,000$. The interest credited in year k is

$$0.08 \min(100000, B) + 0.09 \max(0, B - 100000)$$

where B_k is the beginning year balance. $B_0 = 300,000$, $B_1 = 300,000 + 8,000 + 18,000 - X$, and so on recursively. Use the Goal Seek feature to determine X so that $B_{11} = 0$ (beginning year 11 = end of year 10). $X = 45,797.09$

7. *Guide*: Work from the last year back to the present. The required cash flow for the last year is known and so is the coupon, so you can calculate the number of longest maturity bonds to buy. Then work on the next to the last year, knowing the required cash flow and the number of bonds paying a coupon (but maturing in the following year). And so on.

The total market value is 450,179. You need 1.87 bonds maturing in 1995, etc.

8. *Guide*: Use the Goal Seek feature. The market yield is 7.46

9. *Guide*: Use the Goak Seek feature to find the price for each call date to yield 8%. The price is the minimum of these: 1,085.59.

10. *Guide*: With a trial value for the interest rate, use the future value function (FV) to find the balance after 20 quarters. Use Goal Seek to set the future value to 5000. The solution is $i = 8.58\%$

D.2   The Future Lifetime of a Life Aged x

D.2.1   Solutions to Theory Exercise

1. Use equation (2.2.5). $\mu_{45} = -\frac{d}{dt}\ln(_tp_x)$ evaluated at $t = 45 - x$. Thus

$$
\begin{aligned}
\mu_{45} &= -\frac{d}{dt}\ln\left(\frac{100 - x - t}{100 - x}\right) \\
&= \left.\frac{1}{100 - x - t}\right|_{t=45-x} \\
&= \frac{1}{55}.
\end{aligned}
$$

2. Use equation (2.1.11).

$$
\begin{aligned}
E[T(x)] &= \int_0^\infty {}_tp_x\,dt \\
&= \int_0^{100}\left(1 - \left(\frac{t}{100}\right)^{1.5}\right)dt \\
&= \left.t - \frac{1}{1000}\left(\frac{t^{2.5}}{2.5}\right)\right|_0^{100} \\
&= 60.
\end{aligned}
$$

3. Use (2.2.6). First: $\int_0^{20}\mu_{x+t}\,dt = -\ln(85 - t) - 3\ln(105 - t)|_0^{20} = -\ln\left(\frac{65}{85}\left(\frac{85}{105}\right)^3\right)$. Then ${}_{20}p_x = \frac{65}{85}\left(\frac{85}{105}\right)^3 = 0.4057$.

4. Use (2.1.11).

$$
\begin{aligned}
\overset{\circ}{e}_{41} &= \int_0^\infty {}_tp_{41}\,dt \\
&= \int_0^\infty\left(\frac{42}{42 + t}\right)^3 \\
&= \left.(42)^3\frac{(42 + t)^{-2}}{-2}\right|_0^\infty \\
&= 21.
\end{aligned}
$$

5. The symbol m_x denotes the central death rate: Deaths $d_x = l_x - l_{x+1}$ and average population $= \int_x^{x+1} l_y\,dy = l_x\int_0^1\frac{l_{x+t}}{l_x}\,dt$. Divide each of deaths and average population by l_x to obtain $m_x = \dfrac{q_x}{\int_0^1 {}_tp_x}\,dt$. Use (2.6.9).

$$
\int_0^1 {}_tp_x\,dt = \int_0^1\frac{1 - q_x}{1 - (1 - t)q_x}\,dt
$$

$$= \frac{p_x}{q_x} \ln[1 - (1-t)q_x]\Big|_0^1$$

$$= \frac{-p_x \ln(p_x)}{q_x}.$$

The formula for m_x is the reciprocal of this quantity multiplied by q_x. To work the exercise substitute $q_x = 0.2$ and $p_x = 0.8$. The answer is $m_x = 0.224$.

6. Use equation (2.2.6). $e^{-\mu} = p_x = 1 - 0.16 = 0.84$ so $_tp_x = e^{-t\mu} = (0.84)^t = 0.95$ and, $t = \ln(0.95)/\ln(0.84) = 0.294$.

7. Since $l_x\mu_x$ is constant, l_x is linear. Thus $T(88)$ is uniform on $(0,12)$. Therefore $\mathrm{Var}(T) = (12)^2/12 = 12$.

8. Before: $0.95 = \exp\left(-\int_0^1 \mu_{x+t}dt\right) = p_x$. After: $0.93 = \exp\left(-\int_0^1 (\mu_{x+t} - c)dt\right)$
$= p_x e^c = 0.95 e^c$. Therefore $e^c = 93/95$, $c = \log(93/95) = -0.0213$.

9. Make the change of variables $x + s = y$ in equation (2.2.6) to prove (i). Use the rules for differentiating integrals to prove (ii).

10.

$$100_{1|}q_{[30]+1} = 100\left(p_{[30]+1}\right)\left(q_{[30]+2}\right)$$
$$= \left(1 - q_{[30]+1}\right)(100q_{32})$$
$$= (1 - 0.00574)(0.699)$$
$$= 0.695$$

11.

$$\frac{l_{40} - l_{57}}{l_{21}} = \frac{(81)^{1/2} - (64)^{1/2}}{(100)^{1/2}} = 0.10$$

12. Use this relation:

$$e_x = E[K(x)] = p_x E[K(x)|T(x) \geq 1] + q_x E[K(x)|T(x) < 1] = p_x(1 + e_{x+1}).$$

Thus $p_x = e_x/(1 + e_{x+1})$. $_2p_{75} = p_{75}p_{76} = \frac{10}{1+9.5}\frac{10.5}{1+10.0} = 0.909$.

13. $T(16)$ is uniform on $(0, \omega - 16)$ since we have a de Moivre mortality table. Hence $E[T(16)] = (\omega - 16)/2$ and $\mathrm{Var}[T(16)] = (\omega - 16)^2/12$. Therefore $\omega - 16 = 2(36) = 72$ and $\mathrm{Var}[T] = (72)^2/12 = (72)(6) = 432$.

14. $q_{50} = 1 - {}_{21}p_{30}/{}_{20}p_{30} = 111/6000$. And $\mu_{30+t} = -\frac{1}{{}_tp_{30}}\frac{\partial}{\partial t}{}_tp_{30} = \frac{70+2t}{7800-70t-t^2}$. Therefore, $q_{50} - \mu_{50} = 1/6000$.

15. $E[T] = \int_0^{100-x} {}_tp_x dt = \int_0^a \left(\frac{a-t}{a}\right)^2 dt = \frac{a}{3}$ where $a = 100 - x$. $E[T^2] = \int_0^{100-x} 2t\,{}_tp_x dt = 2\int_0^a t\,{}_tp_x dt$. Use integration by parts to obtain $E[T^2] = \frac{a^2}{6}$. Hence $\mathrm{Var}(T) = E(T^2) - E(T)^2 = a^2(1/6 - 1/9) = (100 - x)^2/18$.

16. $m_x = \frac{q_x}{\int_0^1 {}_tp_x dt}$ and, because of the constant force of mortality, ${}_tp_x = e^{-\mu t}$
where $\mu = -\ln(p_x)$. Hence, $\int_0^1 {}_tp_x dt = q_x/\mu$ and $m_x = \mu = 0.545$.

17. Let $T = T(x)$ be the lifetime of the non-smoker and $T^s = T^s(x)$ be the life-time of the smoker. Use formula (2.2.6): $\Pr(T^s > t) = \exp\left(-\int_0^t c\mu_{x+u}du\right) = ({}_tp_x)^c$ where ${}_tp_x = \Pr(T > t)$. Hence $\Pr(T^s > T) = \int_0^\infty \Pr(T^s > t)g(t)dt = \int_0^\infty ({}_tp_x)^c g(t)dt = -\int_0^\infty w(t)^c w'(t)dt$ where $w(t) = {}_tp_x$. Hence, $\Pr(T^s > T) = \frac{[w(t)]^{c+1}}{c+1}\Big|_0^\infty = 1/(c+1)$.

18. See exercise 9. $q_x = 1 - p_x = 1 - \exp\left(-\int_x^{x+1} \mu_y dy\right)$ which we get by a change of variable of integration in formula (2.2.6). Now apply the rules for differentiation of integrals:

$$\frac{dq_x}{dx} = -\exp\left(-\int_x^{x+1} \mu_y dy\right)(-\mu_{x+1} + \mu_x) = p_x(\mu_{x+1} - \mu_x).$$

19. $\int_{35}^{45} \mu_x dx = 400k$ and so $0.81 = {}_{10}p_{35} = \exp(-\int_{35}^{45} \mu_x dx) = \exp(-400k)$. Similarly ${}_{20}p_{40} = \exp(-\int_{40}^{60} \mu_x dx) = e^{-1000k} = ((0.81)^{1/400})^{1000} = (0.81)^{5/2} = (0.9)^5 = 0.59$

20. $E[X^2] = 2\int_0^\omega x{}_xp_0 dx = 3\omega^2/5$. $\text{Var}[X] = (3\omega^2/5) - (3\omega/4)^2 = \omega^2(3/5 - 9/16) = 3\omega^2/80$.

D.2.2   Solutions to Spreadsheet Exercises

1. See appendix E.

2. Check value: $e_0 = 71.29$.

3. $c = 0.09226$. Assume that "expectation of remaining life" refers to complete life expectation and that *assumption a* applies, so that $\overset{\circ}{e}_x = e_x + 0.5$.

4. Use formula (2.3.4) with $A = 0$. Check values: $\ell_{40} = 99,510$ when $c = 1.01$ and $\ell_{50} = 680$ when $c = 1.20$.

5. Under *assumption a*, $\mu_{x+0.6} = 0.10638$ for example.

6. Under *assumption b*, ${}_{0.4}q_x = 0.04127$ for example.

7. Use trial values such as $B = 0.0001$ and $c = 1$ to calculate Gompertz values, and the sum of their squared differences from the table values. Use the optimization feature to determine values of B and c which minimize the sum. Solution: $B = 2.69210^{-5}$ and $c = 1.105261$.

8. For $k = 7.5$, $e_{45} = 12.924$. For $k = 1$, $e_{45} = 30.890$.

D.3 Life Insurance

D.3.1 Solutions to Theory Exercises

1. The issue age is $x = 30$. From (i), $T = T(30)$ is uniformly distributed on $(0, 70)$. The present value random variable is $Z = 50,000v^T$. Hence, $\bar{A}_{30} = E[Z] = 50,000 \int_0^{70} v^t \frac{1}{70} dt = (50,000/70)(1 - e^{-7})/(0.10) = 7,136$.

2. Use the recursion relation:

$$(IA)_x = A^1_{x:\overline{1}|} + vp_x(A_{x+1} + (IA)_{x+1}).$$

An alternative solution in terms of commutation functions goes like this: The numerator can be written as follows:

$$\frac{R_x - C_x}{D_x} = \frac{M_x + R_{x+1} - C_x}{D_x} = \frac{M_{x+1} + R_{x+1}}{D_x}.$$

The denominator is $\frac{R_{x+1} + M_{x+1}}{D_{x+1}}$. Hence the ratio is $D_{x+1}/D_x = vp_x$.

3. Let Z_3 be the present value random variable for the pure endowment, so $Z_1 = Z_2 + Z_3$. It follows that $\text{Var}(Z_1) = \text{Var}(Z_2) + 2\text{Cov}(Z_2, Z_3) + \text{Var}(Z_3)$. Now use the fact that $Z_2 Z_3 = 0$ to obtain $\text{Cov}(Z_2, Z_3) = -E[Z_2]E[Z_3]$. Z_3 is v^n times the Bernoulli random variable which is 1 with probability ${}_n p_x$, zero otherwise. Hence $\text{Var}(Z_1) = 0.01 + 2(-E[Z_2]E[Z_3]) + \text{Var}(Z_3) = 0.01 - 2(0.04)(0.24) + (0.30)^2(0.8)(1 - 0.8) = 0.0052$.

4. $A_{45:\overline{20}|} = (M_{45} - M_{65} + D_{65})/D_{45} = 0.40822$.

5. Use the recursion relation $A_x = A^1_{x:\overline{n}|} + v^n {}_n p_x A_{x+n}$ and the relation $A_{x:\overline{n}|} = A^1_{x:\overline{n}|} + v^n {}_n p_x$. In terms of the given relations these are $A_x = y + v^n {}_n p_x z$ and $u = y + v^n {}_n p_x$. Hence $A_x = y + (u - y)z = (1 - z)y + uz$.

6. From (ii), the discount function is $v_t = 1/(1 + 0.01t) = 100/(100 + t)$. The benefit function is: $b_t = (10,000 - t^2)/10 = (100 + t)(100 - t)/10$. Hence $Z = v_T b_T = 10(100 - T)$ and so $E[Z] = 10(100 - E[T])$. Now use item (i): $T = T(50)$ is uniform on $(0, 50)$ so $E[T] = 25$ and $E[Z] = 750$.

7. $E[v^T] = \int_0^\infty e^{-\delta t} e^{-\mu t} \mu \, dt = A_x = \frac{\mu}{\mu + \delta}$ and $E[(v^T)^2] = {}^2 A_x = \frac{\mu}{\mu + 2\delta}$.

Therefore, $\text{Var}[v^T] = {}^2 A_x - A_x^2 = \ldots = \frac{\mu \delta^2}{(\mu + 2\delta)(\mu + \delta)^2}$.

8. Consider the recursion relation $A_x = vq_x(1 - A_{x+1}) + vA_{x+1}$. The analog for select mortality with a one year select period goes like this: Since the select period is one year, $K([x] + 1)$ and $K(x + 1)$ are identically distributed. Hence, using the theorem on conditional expectations, we have $A_{[x]} = E[v^{K[x]+1}] = vq_{[x]} + E[v^{K[x+1]+1+1}](1 - q_{[x]}) = vq_{[x]} + E[v^{K(x+1)+1}]v(1 - q_{[x]}) = vq_{[x]} + A_{x+1}v(1 - q_{[x]})$. Hence, $A_{[x]} = vq_{[x]}(1 - A_{x+1}) + vA_{x+1}$. By combining the two recursion relations, we see that $A_x - A_{[x]} = v(q_x - q_{[x]})(1 - A_{x+1}) = 0.5vq_x(1 - A_{x+1})$.

9. Let P be the single premium. Use formula (3.4.3). The benefit function is $c_{k+1} = 10,000 + P$ for $k = 0, 1, \ldots, 19$ and $c_{k+1} = 20,000$ for $k = 20, 21, \ldots$. Hence $PD_x = (10,000 + P)M_x + (10,000 - P)M_{x+20}$. Now solve for $P = 10,000 \dfrac{M_x + M_{x+20}}{D_x - (M_x - M_{x+20})}$.

10. Use formula (3.4.3). Make a column of differences of the death benefit column c_k. In calculating the differences, use a death benefit of 0 at age $x - 1$ and age $x + 11$. Also put in the ages to avoid confusion about which age to use in the solution. You will obtain a table like this:

Age at Death	Year of Death	c_k	$c_k - c_{k-1}$
x	1	10	10
$x+1$	2	10	0
$x+2$	3	9	-1
$x+3$	4	9	0
$x+4$	5	9	0
$x+5$	6	8	-1
$x+6$	7	8	0
$x+7$	8	8	0
$x+8$	9	8	0
$x+9$	10	7	-1
$x+10$	11	0	-7

From the table, we see that the net single premium is written in terms of commutation functions as follows:

$$\frac{10M_x - M_{x+2} - M_{x+5} - M_{x+9} - 7M_{x+10}}{D_x}.$$

11.

$$Z = \begin{cases} v^T & \text{if } T < 10 \\ v^{10} & \text{otherwise} \end{cases}$$

T is uniform on $(0, 50)$. Hence, the net single premium is

$$50,000 \left[v^{10}\,_{10}p_{50} + \int_0^{10} v^t g(t)dt \right] = 50,000 \left[\frac{40}{50} e^{-(0.10)10} + \int_0^{10} v^t \frac{1}{50} dt \right]$$

$$= 10,000[1 + 3e^{-1}] = 21,036.$$

12. Let m be the answer. $m = v^T$, where $_tp_y = 0.5$. Since $_tp_y = \frac{s(y+t)}{s(y)} = e^{-0.02t} = 0.5$, then $m = e^{0.04t} = (e^{-0.02t})^2 = (v.5)^2 = 0.25$.

13. Since $i = 0$, then $Z = 0$ or 1. $Z = 1$ if $K = 0$ or $K = 1$ which occurs with probability $_2q_x = q_x + p_xq_{x+1} = 1/2 + 1/2q_{x+1}$. And $Z = 0$ if $K > 1$, which occurs with probability $1 - (q_x + p_xq_{x+1}) = 1/2 - 1/2q_{x+1}$. Hence Z is Bernoulli;

its variance is $(1/2 + 1/2q_{x+1})(1/2 - 1/2q_{x+1}) = 1/4(1 - q_{x+1}^2)$. Set this equal to 0.1771 and solve. The result is $q_{x+1} = 0.54$.

14. Calculate values of Z and its density function $f(t)$ in the given table. Obtain the following:

t	$c(t)$	q_{x+t}	Z	$f(t)$	$Zf(t)$
0	3	0.20	2.700	0.2	0.5400
1	2	0.25	1.620	0.2	0.3240
2	1	0.50	0.729	0.3	0.2187
≥ 3	0		0	0.3	0

$\Pi = 1.0827$. The values of Z which are greater than $\Pi = 1.0827$ are $Z = 2.7$ and $Z = 1.620$. Hence $\Pr[Z > \Pi] = 0.2 + 0.2 = 0.4$.

15. $A_x = vq_x + vp_x A_{x+1}$ so $A_{76} = vq_{76} + (D_{77}/D_{76})A_{77}$ since $D_{77}/D_{76} = vp_{76}$. Since $p_{76} = (1 + i)(D_{77}/D_{76}) = (1.03)(360/400) = (1.03)(0.9) = 0.927$ then $0.8 = (1.03)^{-1}(1 - 0.927) + (0.9)A_{77}$. Now solve for $A_{77} = 0.810$.

16. The net single premium is $50 \int_0^{100} v^t g(t)dt$. Integrate by parts to get a net single premium of $1 - 11e^{-10} = 0.999501$.

17. See Exercise 7. $E[v^{2T}] = 0.25$ implies that $\frac{\mu}{\mu+2\delta} = 0.25$ so $3\mu = 2\delta$. $E[v^T] = \frac{\mu}{\mu+\delta} = \frac{\mu}{\mu+1.5\mu} = 0.4$.

18. $0.95 = \Pr[(Z_1 + Z_2 + \cdots + Z_{100})1000 \leq 100w]$ where the random variables Z_i are independent and identically distributed like v^T. Now $Y = (1/100)\sum_{k=1}^{100} Z_k$ is approximately normal with mean $E[Z] = 0.06$ and variance equal to $(1/100)\text{Var}(Z) = (1/100)[0.01 - (0.06)^2] = 0.64(10^{-6})$. Thus the mean and standard deviation of Y are 0.06 and 0.008. Therefore $0.95 = \Pr(Y \leq w/1000)$ implies that $w = 1000(0.06 + (1.645)(0.008)) = 73.16$.

19. Use $\Pi = \Pi A^1_{x:\overline{20}|} + 10,000v^{20}{}_{20}p_x$ or $\Pi D_x = \Pi(M_x - M_{x+20}) + 10,000D_{x+20}$ and solve for Π.

20. Use formula (3.4.3). Make a column of differences of the death benefit column c_k. Use a death benefit of 0 at age $x - 1$.

Age at Death	Year of Death	c_k	$c_k - c_{k-1}$
x	1	10	10
$x+1$	2	10	0
$x+2$	3	9	-1
$x+3$	4	9	0
$x+4$	5	9	0
$x+5$	6	8	-1
$x+6$	7	8	0
$x+7$	8	8	0
$x+8$	9	8	0
$x+9$	10	7	-1
$\geq x+10$	≥ 11	7	0

Then the net single premium is given as follows:

$$\frac{10M_x - M_{x+2} - M_{x+5} - M_{x+9}}{D_x}.$$

D.3.2   Solution to Spreadsheet Exercises

1. For $i = 2.5\%, 5.0\%$, and 7.5%, the single premium life insurance at age zero is $A_0 = 0.19629, 0.06463$ and 0.03717.

2. At $i = 5\%$, $(IA)_0 = 2.18345$.

3. Guide: Set up a table with benefits and probabilities of survival to get them. The net single premium is 0.0445.

4. Guide: Use the VLOOKUP() function to construct the array of survial probabilites for a given issue age. Check values: $(D\bar{A})_{50:\overline{50}|} = 9.23$ at $i = 5\%$. $(D\bar{A})_{25:\overline{75}|} = 3.50$ at $i = 6\%$.

5. Guide: Set up a spreadsheet to calculate the values of A_x and ^2A_x. According to formula (3.2.4), the second moment can be calculated by changing the force of interest from δ to 2δ. Put δ in a cell and let it drive the interest calculations. Use the Data Table (or What if?) feature to find the two values of $\mathrm{E}[v^{K+1}]$, corresponding to δ and 2δ. Check values: $\mathrm{Var}(v^{K+1}) = 20,190$ when $i = 5\%$, and $\mathrm{Var}(v^{K+1}) = 17,175$ when $i = 2\%$.

D.4 Life Annuities

D.4.1 Solutions to Theory Exercises

1. Use formula (4.3.9) with $m = 2, x = 40, n = 30$. $\ddot{a}_{40:\overline{30}|} = D_{40}^{-1}(N_{40} - N_{70}) = 15.1404$, $D_{70}/D_{40} = v^{30}p_{40} = 0.1644$, $\alpha(2) = \frac{id}{i^{(2)}d^{(2)}} = 1.00015$, and $\beta(2) = 0.25617$. The answer is $\ddot{a}_{40:\overline{30}|}^{(2)} = 14.9286$.

2. Use the recursion formula (4.6.2) for $a_{x:\overline{1}|}$ to develop the following recursion formula: $(I\ddot{a})_x = \ddot{a}_{x:\overline{1}|} + vp_x(\ddot{a}_{x+1} + (I\ddot{a})_{x+1})$. The ratio simplifies consequently to $vp_x = a_{x:\overline{1}|}$.

3. $(\bar{I}_{\overline{n}|}\bar{a})_x = \int_0^n tv^t{}_tp_x dt + \int_n^\infty nv^t{}_tp_x dt$. Differentiate before making the substitutions $v_t = e^{-0.06t}$ and ${}_tp_x = e^{-0.04t}$. Use Liebnitz's rule for differentiating integrals:

$$\frac{\partial}{\partial n}(\bar{I}_{\overline{n}|}\bar{a})_x = nv^n{}_np_x + \int_n^\infty v^t{}_tp_x dt - nv^n{}_np_x$$
$$= \int_n^\infty v^t{}_tp_x dt = \int_n^\infty e^{-0.10t} dt = 10e^{-0.10n}.$$

4. Arrange the calculations in a table:

Event	Pr[Event]	Present Value (PV)	(PV)Pr[Event]	(PV)^{2}Pr[Event]
$K = 0$	0.2	2.00	0.400	0.800
$K = 1$	0.2	4.70	0.940	4.418
$K \geq 2$	0.6	7.94	4.764	37.826

$E[PV] = 6.104$ and $E[PV^2] = 43.044$. Hence, the variance is $43.044 - (6.104)^2 = 5.785$.

5. $(I\bar{a})_{95} = \bar{a}_{95} + {}_{1|}\bar{a}_{95} + {}_{2|}\bar{a}_{95} + {}_{3|}\bar{a}_{95} + \cdots$. Since $\omega = 100$, $i = 0$ and $T(95)$ is uniform on $(0, 5)$, then the five non-zero terms are $\bar{a}_{95} = E[T(95)] = 2.5, {}_{1|}\bar{a}_{95} = p_{95}\bar{a}_{96} = (0.8)(2) = 1.6, {}_{2|}\bar{a}_{95} = {}_2p_{95}\bar{a}_{97} = (0.6)(1.5) = 0.9, {}_{3|}\bar{a}_{95} = {}_3p_{95}\bar{a}_{98} = (0.4)(1) = 0.4$ and ${}_{4|}\bar{a}_{95} = {}_4p_{95}\bar{a}_{99} = (0.2)(0.5) = 0.1$. Hence, the answer is 5.5. Alternatively, we can calculate expected present values conditionally on the year of death. There are five years of interest and they are equally likely. This yields $(0.5 + 2.0 + 4.5 + 8.0 + 12.5)/5 = 5.5$.

6. Use formula (4.3.9) or its equivalent in terms of commutation functions. ${}_{10|}\ddot{a}_{25:\overline{10}|} = D_{25}^{-1}(N_{35} - N_{45}) = 4.85456$, $(D_{35} - D_{45})/D_{25} = 0.24355$, $\alpha(12) = 1.00020$, and $\beta(12) = 0.46651$. Hence,

$${}_{10|}\ddot{a}_{25:\overline{10}|}^{(12)} = \alpha(12)\,{}_{10|}\ddot{a}_{25:\overline{10}|} - \beta(12)({}_{10}p_{25}v^{10} - {}_{20}p_{25}v^{20}) = 4.74191.$$

Another solution is based on formula (4.3.2) and (3.3.10), adjusted for temporary rather than whole life contracts:

$$A^{(12)}_{35:\overline{10}|} = \frac{i}{i^{(12)}} A^1_{35:\overline{10}|} + {}_{10}p_{35}v^{10}$$

$$= \frac{i}{i^{(12)}} (M_{35} - M_{45})/D_{35} + D_{45}/D_{35} = 0.61814.$$

Hence,

$$_{10|}\ddot{a}^{(12)}_{25:\overline{10}|} = (D_{35}/D_{25}) \left(1 - A^{(12)}_{35:\overline{10}|}\right)/d^{(12)} = 4.74200.$$

7. Use formula (4.3.9) to derive a formula analogous to formula (4.5.4) for temporary annuities. Then use the formulas analogous to (A.3.6) and (A.3.9) to write the result in terms of commutation functions:

$$(I\ddot{a})^{(m)}_{x:\overline{n}|} = \alpha(m)\left((I\ddot{a})_{x:\overline{n}|}\right) - \beta(m)\left(\ddot{a}_{x:\overline{n}|} - nv^n{}_np_x\right)$$

$$= \alpha(m)\frac{S_x - S_{x+n} - nN_{x+n}}{D_x} - \beta(m)\frac{N_x - N_{x+n} - nD_{x+n}}{D_x}$$

Now calculate the N and D values by differencing the successive values of the given values of S. We need $N_{70} = S_{70} - S_{71} = 9597$, $N_{80} = S_{80} - S_{81} = 2184$, $D_{70} = N_{70} - N_{71} = 9597 - 8477 = 1120$, and $D_{80} = N_{80} - N_{81} = 368$. We get $(I\ddot{a})^{(12)}_{70:\overline{10}|} = 29.16$.

8.

$$_np_x d\ddot{a}_{\overline{n}|} + d\sum_{k=0}^{n-1}\frac{(1 - v^{k+1})}{d}\,{}_kp_xq_{x+k}$$

$$= d\left(\ddot{a}_{\overline{n}|}\Pr(K \geq n) + \sum_{k=0}^{n-1}\ddot{a}_{\overline{k+1}|}\Pr(K = k)\right)$$

$$= dE\left(\ddot{a}_{\overline{\min(K+1,n)}|}\right)$$

$$= d\ddot{a}_{x:\overline{n}|} = 1 - A_{x:\overline{n}|}$$

9. Use formulas (4.2.9), (3.2.4) and (3.2.5). $\text{Var}(Y) = d^{-2}\left(\text{E}[e^{-2\delta(K+1)}] - A_x^2\right)$. Now use $A_x = 1 - d\ddot{a}_x$ twice. A_x evaluated at δ is $1 - (0.04)10 = 0.6$. The discount corresponding to 2δ is $1 - v^2 = 1 - (0.96)^2 = 0.0784$ so $\text{E}[e^{-2\delta(K+1)}] =$ "A_x evaluated at 2δ," is $1 - (0.0784)(6) = 0.5296$. Therefore and $\text{Var}(Y) = (0.5296 - (0.6)^2)/(0.04)^2 = 106$.

10. Use formulas (4.2.13) and (3.2.12).

11. Use formula (A.4.7). $N_{28} = S_{28} - S_{29} = 97$, $N_{29} = S_{29} - S_{30} = 93$, and $D_{28} = N_{28} - N_{29} = 4$. Hence, $M_{28} = 4 - (3/103)97 = 1.1748$ where we used the commutation function version of formula (4.2.8): $D_x = dN_x + M_x$.

12. Use the recursion relation $\ddot{a}_x = 1 + vp_x\ddot{a}_{x+1}$. $7.73 = 1 + (1.03)^{-1}p_{73}(7.43)$ so $p_{73} = (1.03)(6.73)/7.43 = 0.93$.

13. The values of the present value random variable Y are 2, $2 + 3v = 4.7273$, and $2 + 3v + 4v^2 = 8.0331$. Hence, $\Pr(Y > 4) = \Pr(K > 0) = 1 - 0.20 = 0.80$

14. Use formula (4.4.8) with $r(t) = 1$ if $0 \leq t < 1$ and $r(t) = 2$ for $t \geq 1$. Integration by parts applied to $E(Y) = \int_0^{20} r(t)(1 - 0.05t)dt$ with $w(t) = \int_0^t r(s)ds$ yields $E(Y) = 0.05 \int_0^{20} w(t)dt = 19.025$. Alternatively, the annuity can be viewed as the sum of two annuities each having constant rate of payment of 1 per year. The first begins paying at age 80, the second at age 81. Using this approach we have $E(Y) = \bar{a}_{80} + vp_{80}\bar{a}_{81} = E(T(80)) + 0.95E(T(81))$ where $T(80)$ and $T(81)$ are uniformly distributed on $(0, 20)$ and $(0, 19)$, respectively. Again we get $E(Y) = 10 + 0.95(9.5) = 19.025$.

15. Consider the sum of two annuities approach, as in exercise 14: $E(Y) = \bar{a}_{80:\overline{5}|} + vp_{80}\bar{a}_{81:\overline{4}|} = E[\min(T(80), 5)] + 0.95E[\min(T(81), 4)] = (2.5)(0.25) + (5)(0.75) + (0.95)[(2)(4/19) + 4(15/19)] = 7.775$.

16. Since $\delta = 0$, $h = \text{Var}(T) = E(T^2) - (E(T))^2$. Also $E(T^2) = \int_0^\infty t^2 g(t)dt = 2\int_0^\infty t(1 - G(t))dt = 2g$ by parts integration. Hence, $E(T) = \sqrt{2g - h}$.

17. $(D\ddot{a})_{70:\overline{10}|} = D_{70}^{-1}(10N_{70} - S_{71} + S_{81}) = 42.09$.

18.

$$\frac{a_{\overline{1}|}S_x - \ddot{a}_{\overline{2}|}S_{x+1} + \ddot{a}_{\overline{1}|}S_{x+2}}{D_x} = \frac{vS_x - (1+v)S_{x+1} + S_{x+2}}{D_x}$$

$$= \frac{vN_x - N_{x+1}}{D_x}.$$

The formula (A.4.6) $C_y = vD_y - D_{y+1}$, summed over y running from x to the end of the table, gives $M_x = vN_x - N_{x+1}$, from which we see the simplification to A_x.

19. Let $Z = e^{-\delta T}$ and $Y = \bar{a}_{\overline{T}|} = \delta^{-1}(1 - Z)$. From the given data, we find that $E(Z) = 1 - 10\delta$, $E(Z^2) = 1 - 14.75\delta$, and $50 = \text{Var}(Y) = \delta^{-2}\left(E(Z^2) - E(Z)^2\right)$. First solve for $\delta = 3.5\%$. Then $\bar{A}_x = 1 - \delta\bar{a}_x = 0.65$.

20. Apply formula (4.8.9) to obtain $A_{35.75} = 0.17509$. Apply formula (3.3.5) to obtain $\bar{A}_{35.75} = 0.18046$. Then $\bar{a}_{35.75} = (1 - \bar{A}_{35.75})/\delta = 16.79725$.

D.4.2 Solutions to Spreadsheet Exercises

1. Set up your spreadsheet to calculate the required annuity value with reference to a single age and interest rate. Use VLOOKUP() references to the mortality values, which may be on a separate sheet. Then use the Data Table feature to calculate the array of values for x running down a column and i accross a row. Check values: $\ddot{a}_{30} = 18.058$ at $i = 5\%$ and $\ddot{a}_{20} = 13.753$ at $i = 7.5\%$.

2. The expected market value is 309,153.

3. Guide: Set up a table to calculate A_x and $E[Z^2]$ with a reference to a single value of c. Use formula (3.2.4). Then use the Table feature to allow for different values of c.

4. For $i = 5\%$, $\ddot{a}^{(2)}_{24.5} = 18.3831$ and $A_{24.5} = 0.11255$. For $i = 6\%$, $\ddot{a}^{(12)}_{30.25} = 15.37108$ and $A_{30.25} = 0.10369$.

5. Guide: From the Illustrative Life Table set up a table with the cummulative distribution function of K. Fill a column with 200 random numbers from the interval $[0, 1]$ using RAND(). Use the VLOOKUP() function to find the corresponding value of K. Evaluate Y for each value of K, then calculate the sample mean and variance using the built-in functions. The theoretical answers are $\ddot{a}_{40} = 16.632$ and $\mathrm{Var}(Y) = 10.65022$.

D.5 Net Premiums:Solutions

D.5.1 Theory Exercises

1. Use formula (5.3.15): $P_{25:\overline{20}|} = P^1_{25:\overline{20}|} + P_{25:\frac{1}{20}|} = 0.064$. Now use the relation $_{20}P_{25} = P^1_{25:\overline{20}|} + P_{25:\frac{1}{20}|} A_{45} = 0.046$ to obtain

$$P_{25:\frac{1}{20}|} = (0.064 - 0.046)/(1 - 0.64) = 0.05.$$

Therefore $P^1_{25:\overline{20}|} = 0.064 - 0.05 = 0.014$.

2. $L^* = v^{K+1} - G\ddot{a}_{\overline{K+1}|} = -G/d + (1 + G/d)v^{K+1}$ and hence

$$\text{Var}[L^*] = (1 + G/d)^2 \text{Var}[v^{K+1}] = (d + G)^2 d^{-2} \text{Var}[v^{K+1}].$$

Similarly,

$$L = v^{K+1} - P_x \ddot{a}_{\overline{K+1}|} = -P_x/d + (1 + P_x/d)v^{K+1}$$

and

$$\text{Var}[L] = (d + P_x)^2 d^{-2} \text{Var}[v^{K+1}] = 0.30.$$

Now use $\text{E}[L] = 0$ and $\text{E}[L^*] = -0.20$ to find that $0 = A_x - P_x \ddot{a}_x = 1 - (d + P_x)\ddot{a}_x$ and $-0.2 = 1 - (d + G)\ddot{a}_x$. Hence

$$\begin{aligned}
\text{Var}[L^*] &= (d + G)^2 d^{-2} \text{Var}[v^{K+1}] = 0.30(d + G)^2/(d + P_x)^2 \\
&= 0.30[(d + G)/(d + P_x)]^2 = 0.432.
\end{aligned}$$

3. Let P denote the net annual premium.

$$\begin{aligned}
P &= \frac{5.000(10A_{25} + {}_{5|}A_{25} + {}_{10|}A_{25} + {}_{15|}A_{25} + {}_{20|}A_{25} + {}_{25|}A_{25}}{\ddot{a}_{25:\overline{10}|}} \\
&= \frac{5.000(10M_{25} + M_{30} + M_{35} + M_{40} + M_{45} + M_{50}}{N_{25} - N_{35}} \\
&= 1012.33.
\end{aligned}$$

4.

$$\begin{aligned}
\text{Loss A} &= 4v^{K+1} - 0.18\ddot{a}_{\overline{K+1}|} \\
&= -0.18/d + (4 + 0.18/d)v^{K+1} = -2.25 + 6.25v^{K+1}
\end{aligned}$$

Using the table we find that $\text{Var}[\text{Loss A}] = 3.25 = (6.25)^2 \text{Var}[v^{K+1}]$. Similarly, $\text{Loss B} = 6v^{K+1} - 0.22\ddot{a}_{\overline{K+1}|} = -2.75 + 8.75v^{K+1}$ and $\text{Var}[\text{Loss B}] = (8.75)^2 \text{Var}[v^{K+1}] = (8.75)^2(3.25)/(6.25)^2 = 6.37$.

5. Let Q be the answer. The death benefit is $10,000 + Q/2$ initially. After the premiums are paid up it reduces to 10,000. Hence, the expected value of the present value of benefits is $10,000A_x + (Q/2)\,A^1_{x:\overline{20}|}$. The expected present value of premiums is $Q\ddot{a}_{x:\overline{20}|}$. The equivalence principle implies that

$$Q = \frac{10,000A_x}{\ddot{a}_{x:\overline{20}|} - (A^1_{x:\overline{20}|})/2}.$$

Now rearrange the denominator.

$$\ddot{a}_{x:\overline{20}|} - (A^1_{x:\overline{20}|})/2 = \ddot{a}_{x:\overline{20}|} - (1 - d\ddot{a}_{x:\overline{20}|} - A_{x:\frac{1}{20}|})/2$$
$$= (1 + d/2)\ddot{a}_{x:\overline{20}|} - (1 - v^{20}{}_{20}P_x)/2.$$

Hence,

$$Q = \frac{10,000A_x}{(1 + d/2)\ddot{a}_{x:\overline{20}|} - (1 - v^{20}{}_{20}P_x)/2}.$$

6. $A_x = {}_nP_x\ddot{a}_{x:\overline{n}|} = {}_nP_x(1 - A_{x:\overline{n}|})/d$. Hence ${}_nP_x = dA_x/(1 - A_{x:\overline{n}|})$.

7. The present value of death benefits is $1000(1.06)v$ in year 1, $10000(1.06)^2v^2$ in year 2, *etc.* Since $v = 1/1.06$, the present value of death benefits is 1000, independent of the year of death. Let Q be the net annual premium. The expected present value of net annual premiums is $Q\ddot{a}_x$. Hence $Q = 1,000/\ddot{a}_x = 1000(d + P_x) = 1000(0.06/1.06) + 1000P_x = 66.60$.

8. Solve the two equations:

$A_x - A^1_{x:\overline{20}|} + v^{20}{}_{20}p_xA_{x+20}$ and $A_{x:\overline{20}|} - A^1_{x:\overline{20}|} = v^{20}{}_{20}p_x$ for $v^{20}{}_{20}p_x = 0.5$ and $A^1_{x:\overline{20}|} = 0.05$.

Use $A_{x:\overline{20}|} = 1 - d\ddot{a}_{x:\overline{20}|}$ to find $\ddot{a}_{x:\overline{20}|} = 15.45$. This yields $1000P\left(\bar{A}_{x:\overline{20}|}\right) = 1000\left(\bar{A}^1_{x:\overline{20}|} + v^{20}{}_{20}p_x\right)/\ddot{a}_{x:\overline{20}|} = 1000\left[(\frac{i}{\delta})A^1_{x:\overline{20}|} + v^{20}{}_{20}p_x\right]/\ddot{a}_{x:\overline{20}|} = 35.65$

9. $L = v^T - \bar{P}(\bar{A}_x) = -\bar{P}(\bar{A}_x)/\delta + (\delta + \bar{P}(\bar{A}_x))(\delta)^{-1}v^T$. By the continuous payment analog of formula (5.3.5), we have $\bar{a}_x(\delta + \bar{P}(\bar{A}_x)) = 1$. Hence, $\mathrm{Var}(L) = \mathrm{Var}(v^T)/(\delta\bar{a}_x)^2$. By exercise 7 of chapter 3,

$$\mathrm{Var}[v^T] = E[v^{2T}] - (E[v^T])^2 = \frac{\mu\delta^2}{(2\delta + \mu)(\delta + \mu)^2}.$$

Finally, $\mathrm{Var}(L) = \mu/(2\delta + \mu)$ since $\bar{a}_x = 1/(\delta + \mu)$.

10. (a) The insurer's loss random variable is the present value of benefits less the present value of premiums. The death benefit payable at the moment of death is $1 + G\bar{s}_{\overline{T}|}$, provided $T < 10$. The present value of death benefits is

$v^T(1 + G\ddot{s}_{\overline{T}|}) = v^T + G\bar{a}_{\overline{T}|}$ if $T < 10$ and 0 if $T \geq 10$. The present value of premiums is $G\bar{a}_{\overline{T}|}$ if $T < 10$ and $G\bar{a}_{\overline{10}|}$ if $T \geq 10$. Therefore, the insurer's loss is

$$L = \begin{cases} v^T & \text{if } T < 10 \\ -G\bar{a}_{\overline{10}|} & \text{otherwise.} \end{cases}$$

(b) $E[L] = E[v^T | T < 10]_{10}q_x + (-G\bar{a}_{\overline{10}|})_{10}p_x = \bar{A}^1_{x:\overline{10}|} - G\bar{a}_{\overline{10}|} {}_{10}p_x.$

$E[L^2] = E[v^{2T}|T < 10]_{10}q_x + (-G\bar{a}_{\overline{10}|})^2 {}_{10}p_x = {}^2\bar{A}^1_{x:\overline{10}|} + G^2(\bar{a}_{\overline{10}|})^2 {}_{10}p_x.$

$\text{Var}[L] = {}^2\bar{A}^1_{x:\overline{10}|} + (G\bar{a}_{\overline{10}|})^2 {}_{10}p_x - (\bar{A}^1_{x:\overline{10}|} - G\bar{a}_{\overline{10}|} {}_{10}p_x)^2.$

(c) If G is determined by the equivalence principle, then $G = \bar{A}^1_{x:\overline{10}|}/(\bar{a}_{\overline{10}|} {}_{10}p_x)$. This can be substituted into the last expression to find

$\text{Var}[L] = {}^2\bar{A}^1_{x:\overline{10}|} + (\bar{A}^1_{x:\overline{10}|})^2/{}_{10}p_x.$

11. $\ddot{a}_{30:\overline{10}|} = 1 + a_{30:\overline{9}|} = 6.6.$ Hence $A_{30:\overline{10}|} = 1 - d\ddot{a}_{30:\overline{10}|} = 0.4$ Therefore $A^1_{30:\overline{10}|} = A_{30:\overline{10}|} - v^{10}{}_{10}p_{30} = 0.4 - 0.035 = 0.05.$ Hence, $1000P^1_{30:\overline{10}|} = 1000A^1_{30:\overline{10}|}/\ddot{a}_{30:\overline{10}|} = 7.58.$

12. $A_x = 0.2$ and $\bar{A}_x = (i/\delta)A_x = 0.2049593$ by *assumption a.* Hence, $\ddot{a}_x = (1 - A_x)/d = 16.8$ and hence $10000P(\bar{A}_x) = 10000\bar{A}_x/\ddot{a}_x = 2,049.59/16.8 = 122.$ Also, $\bar{a}_x = (1 - \bar{A}_x)/\delta = 16.2951.$ Hence, $10000\bar{P}(\bar{A}_x) = 2049.59/16.2951 = 125.78$ and the answer is 3.78.

13.

$$1 - \frac{\left(P_{30:\overline{15}|} - {}_{15}P_{30}\right)\ddot{a}_{30:\overline{15}|}}{v^{15}({}_{15}P_{30})} = 1 - \frac{A_{30:\overline{15}|} - A_{30}}{{}_{15}E_{30}}$$

$$= 1 - \frac{A_{30:\overline{15}|} - A_{30}}{A^{\;\;1}_{30:\overline{15}|}}$$

$$= \frac{A^{\;\;1}_{30:\overline{15}|} - A_{30:\overline{15}|} + A_{30}}{A^{\;\;1}_{30:\overline{15}|}}$$

$$= \frac{A_{30} - A^1_{30:\overline{15}|}}{A^{\;\;1}_{30:\overline{15}|}}$$

$$= \frac{M_{30} - (M_{30} - M_{45})}{D_{45}} = A_{45}.$$

14. Under *assumption a*, $A_x = \frac{\delta}{i}\bar{A}_x = \frac{0.07}{e^{0.07} - 1}0.03 = 0.289622.$

$$\ddot{a}_x = \frac{1 - Ax}{d} = 10.507587$$

Using (4.3.5), $\ddot{a}_x^{(2)} = \alpha^{(2)}\ddot{a}_x - \beta^{(2)} = 10.252457$.

Therefore, $1000\left(0.5P^{(2)}(\bar{A}_x)\right) = \dfrac{500\bar{A}_x}{\ddot{a}_x^{(2)}} = 14.63$.

15. $\text{Var}[L] = 0.25$ where $L = v^T - \bar{P}\left(\bar{A}_x\right)\bar{a}_{\overline{T}|} = (\delta\bar{a}_x)^{-1}v^T - \bar{P}\left(\bar{A}_x\right)/\delta$.
Therefore $\text{Var}[L] = \text{Var}[v^T]/(\delta\bar{a}_x)^2 = (100/36)\text{Var}[v^T]$. Hence $0.25 = \text{Var}[L] = (100/36)\text{Var}[v^T]$ and $\text{Var}[v^T] = 0.09$. $L^* = v^T - G\bar{a}_{\overline{T}|} = (1 + G/\delta)v^T - G/\delta$
and $\text{Var}[L^*] = (11/6)^2\text{Var}[v^T]$ and hence $\text{Var}[L^*] = (11/6)^2\text{Var}[v^T] = 0.3025$.

16. Let $P = \dfrac{A_{40:\overline{20}|}}{k}$ = net annual premium. Then $P\ddot{a}_{40:\overline{20}|} = A_{40:\overline{20}|} + E[W]$ where $W = Pv^{K+1}\ddot{s}_{\overline{K+1}|}$ if $K < 10$ and $W = 0$ if $K \geq 10$. Since
$E[W] = P\left(\ddot{a}_{40:\overline{10}|} - {}_{10}E_{40}\ddot{s}_{\overline{10}|}\right)$, $P\left(\ddot{a}_{40:\overline{20}|} - \ddot{a}_{40:\overline{10}|} + {}_{10}E_{40}\ddot{s}_{\overline{10}|}\right) = A_{40:\overline{20}|}$
and therefore

$$\begin{aligned}
k &= \ddot{a}_{40:\overline{20}|} - \ddot{a}_{40:\overline{10}|} + {}_{10}E_{40}\ddot{s}_{\overline{10}|} \\
&= \left(\ddot{a}_{50:\overline{10}|} + \ddot{s}_{\overline{10}|}\right){}_{10}E_{40}.
\end{aligned}$$

17. Let P denote the annual premium. Then $L = -P - Pv$ with probability $(0.9)(0.8) = 0.72, L = v^2 - P - Pv$ with probability $(0.9)(0.2) = 0.18$ and $L = v - P$ with probability 0.1. By the equivalence principle, $P = \left(vq_x + v^2p_xq_{x+1}\right)/(1 + vp_x) = 0.13027621$. Thus the values of L are $L = -0.2475$ with probability 0.72. $L = 0.5625$ with probability 0.18 and $L = 0.7697$ with probability 0.1. Hence $\text{Var}[L] = E[L^2] = (-0.2475)^2(0.72) + (0.5625)^2(0.18) + (0.7697)^2(0.1) = 0.160$

18. $1,000\bar{P}\left(\bar{A}_{x:\overline{n}|}\right) = 1,000\bar{A}_{x:\overline{n}|}/\bar{a}_{x:\overline{n}|}$. $0.4275 = \bar{A}^1_{x:\overline{n}|} = \int_0^n \mu e^{-(\delta+\mu)t}dt = 0.45\left(1 - e^{-0.1n}\right)$ and hence $e^{-0.1n} = 0.05$. Therefore $\bar{A}_{x:\overline{n}|} = 0.05 + 0.4275 = 0.4775$ and $\bar{a}_{x:\overline{n}|} = (1 - 0.4775)/0.055 = 9.5$. $1,000\bar{P}\left(\bar{A}_{x:\overline{n}|}\right) = 1,000(0.4775)/9.5 = 50.26$.

19. The loan payment for a loan of $1,000$ is $P = 1,000/a_{\overline{4}|} = 288.60$. The death benefit paid at $K+1$ is $b_{K+1} = P\ddot{a}_{\overline{4-K}|}$ if $K < 4$ and $b_{K+1} = 0$ otherwise. The present value of the death benefit is $Z = v^{K+1}b_{K+1} = Pv^{K+1}\left(1 - v^{4-K}\right)/d$ if $K < 4$ and $Z = 0$ if $K \geq 4$.

(a) $L = Z - G = P\left(v^{K+1} - v^5\right)/d - G$ if $K < 4$ and $L = -G$ if $K \geq 4$.

(b) $0 = E[L] = E[Z] - G = P\left(A^1_{25:\overline{4}|} - v^5{}_4q_{25}\right)/d - G$.

So now calculate as follows: $A^1_{25:\overline{4}|} = A_{25:\overline{4}|} - v^4{}_4p_{25} = 1 - d\ddot{a}_{25:\overline{4}|} - v^4{}_4p_{25} = 0.0043$. $G = 288.6(0.0043 - 0.00373629)/d = 2.87$.

(c)   Let G^* denote the additional amount of the borrowing to pay for term insurance and L^* denote the loss random variable in this case.

$$L^* = \begin{cases} \frac{10,000+G^*}{a\,\overline{4|}}(v^{k+1} - v^5)\frac{1}{d} - G^* & (k < 4) \\ -G^* & \text{(otherwise)} \end{cases}$$

$$E[L^*] = \frac{10,000 + G^*}{a_{\overline{4|}}\,d}(A^1_{25:\,\overline{4|}} - v^5{}_4q_{25}) - G^* = 0$$

Using the result in (ii), $A^1_{25:\,\overline{4|}} - v^5{}_4q_{25} = 0.0005638$. Therefore

$(10,000 + G^*)(0.0028745) - G^* = 0$, and $G^* = 28.827866$. The annual payments is $(10,000+28,827866)/a_{\overline{4|}} = 2,894.23$.

20. (a) $L = v^{K+1} - G\ddot{a}_{\overline{K+1|}} = (1 + G/d)v^{K+1} - G/d$ where K is the curtate lifetime of (x).

(b) $E[L] = A_x - G\ddot{a}_x = (1 + G/d)A_x - G/d = -0.08$ and
$\text{Var}[L] = (1 + G/d)^2({}^2A_x - A_x^2) = 0.0496$.

(c) $0.05 = $ probability of loss $= Pr[S > 0]$ where $S = L_1 + \ldots + L_N$ and the L_i are independent and distributed like L. Thus $E[S] = NE[L] = N(-0.08)$ and $\text{Var}[S] = N\text{Var}[L] = N(0.0496)$ and, using the normal approximation, we have $0.95 = Pr[S \le 0] = Pr(Z \le (0 - N(-0.08))/(0.0496N)^{1/2}) = Pr(Z \le N^{1/2}(0.3592))$. This gives $1.645 = N^{1/2}(0.3592)$ and $N = 20.97$. So a portfolio of $N = 21$ would have a probability of a loss of a little less than 0.05.

D.5.2   Solutions to Spreadsheet Exercises

1. Check value: $S_0 = 21,834,463$.

2. For $i = 5\%$, the premium is 0.0253. For $i = 8\%$, it is 0.0138.

Guide: Set up a spreadsheet with survival probabilities, increasing benefits and increasing premiums. Calculate the expected present value of benefits and premiums and use the solver feature to find the initial premium so that the difference is 0.

3. Some values of the premium P are as follows:

$P = 36675.49$ with $C = 500,000$, $a = 10^{-6}$, and

$P = 43920.03$ with $C = 500,000$, $a = 10^{-5}$,

$P = 375046$ with $C = 1,000,000$, $a = 10^{-5}$.

Guide: Set up a table with columns for the present value of benefits and present value of premiums for each of the ten years. Find L and $U(-L)$ and calculate $E[U(-L)]$. Use the solver feature to find the premium so that formula (5.2.9) holds.

4. P=3.0807.

Guide: Set up a table with probabilities of survival, death benefits and premiums. Set the death benefits according to the refund condition. Set the expected present value of premiums and benefits equal by using the solver function and letting the premium vary. Try it with different death probabilities for the first five years.

5. Guide: Set up a spreadsheet with survival probabilities and present value of premiums. Find the balance and present the value of benefits. Use the solver feature to find the premium that equates the present value of premiums to the present value of benefits.

6. $z = 8.5\%$ for $j = 3\%$ and $z = 14.3\%$ for $j = 6\%$

Guide: Set up a spreadsheet with survival probabilities and cash flows. Cash flows consist of savings during the pre-retirement period and payments during the retirement period. Use the solver feature to find z so that the expected NPV of the cash flows at 5% is 0.

7. The ratio is 1.0601.

D.6 Net Premium Reserves:Solutions

D.6.1 Theory Exercises

1. $P_{35:\overline{20}|} = 0.03067, P_{40:\overline{15}|} = 0.04631$, Reduced Paid-up $= 337.84$.

2. Use the formula

$$
\begin{aligned}
1 - {}_{20}V_{25} &= \frac{\ddot{a}_{45}}{\ddot{a}_{25}} \\
&= \frac{\ddot{a}_{35}}{\ddot{a}_{25}} \frac{\ddot{a}_{45}}{\ddot{a}_{35}} \\
&= (1 - {}_{10}V_{25})(1 - {}_{10}V_{35}) = 0.72
\end{aligned}
$$

and ${}_{20}V_{25} = 0.28$.

3. ${}_{20}\bar{V}\left(\bar{A}_{40}\right) - {}_{20}V\left(\bar{A}_{40}\right) = 1 - \dfrac{\bar{a}_{60}}{\bar{a}_{40}} - {}_{20}V\left(\bar{A}_{40}\right) = 1 - \dfrac{12.25}{20} - 0.3847 = 0.0028$.

4.

k	0	1	2
${}_{k+1}V$	0.8889	1.7984	2.2187.

5. $\ddot{a}_{x+1} = 1.14151, \ddot{a}_x = 1.37691$, Answer: 0.171.

6. Answer: 258.31

7. Use

$$
\begin{aligned}
\text{Var}(L) &= \text{Var}\left[v^{K+1} - P_x\frac{1 - v^{K+1}}{d}\right] \\
&= (1 + P_x/d)^2 \, \text{Var}\left[v^{K+1}\right] \\
&= \frac{1}{(1 - A_x)^2}\left(\text{E}[e^{-(K+1)2\delta}] - \left(\text{E}[e^{-(K+1)\delta}]\right)^2\right)
\end{aligned}
$$

Calculate the moment generating function of $K + 1$,

$$
M(-s) = A_x = \sum_{k=0}^{\infty} e^{-(k+1)s}\,{}_{k|}q_x = \sum_{k=0}^{\infty} v^{k+1}(0.5)^{k+1} = \frac{v(0.5)}{1 - v(0.5)} = \frac{v}{2 - v}
$$

where $v = e^{-s}$. Obtain $E\left[e^{-(K+1)\delta}\right]$ by setting $s = \delta$, and $E\left[e^{-(K+1)2\delta}\right]$ by setting $s = 2\delta$. Substitute and simplify. It is not easy.

8. Answer: 4.88.

9. Answer: 480.95.

10. Answer 644.50.

11. Premium $= 7.92$, Reserve $= 33.72$.

12. $\Lambda_1 = \begin{cases} 0 & \text{with probability } 0.2 \\ 3v - {}_1V - 1 & \text{with probability } 0.2 \\ {}_2Vv - {}_1V - 1 & \text{with probability } 0.6 \end{cases}$

$E[\Lambda_1] = 0$, and $\text{Var}[\Lambda_1] = 0.1754$

13. Answer: 0.2841.

14. Use formula (6.7.9): $\text{Var}[\Lambda_9] = v^2(1.000)^2(1 - {}_{10}V_{40})^2 {}_9p_{40}p_{49}q_{49} =$

$(\dfrac{1}{1.05})^2(1.000)^2(14.63606/16.632258)^2 \dfrac{8,950,994}{9,313,144}0.00546 = 3,685.83$

15. The recursion relation between the two reserves is especially simple since we are beyond the premium paying period: $0.585(1.04) = 0.600p_{38} + q_{38}$ and hence $p_{38} = [1 - (1.04)(0.585)] \div (0.4) = 0.979$.

16. Use ${}_tV_x = 1 - (P_x + d)\ddot{a}_{x+t}$ to solve for $d = \frac{1}{11}$ and $i = 0.10$.

17. Answer: 15.25.

18. Answer: 3.99.

19. (a) $1000{}_{10.5}V_x = 0.5(311 + 340.86) + 0.5(60) = 355.93$.

(b)

$$1000{}_{10.5}V_x = 1000(v^{0.5}{}_{0.5}p_{x+10+0.5}\,({}_{11}V_x)) + 1000v^{0.5}{}_{0.5}q_{x+10+0.5}$$

Use *assumption a* to obtain ${}_{0.5}q_{x+10+0.5} = \dfrac{0.5q_{x+10}}{1 - 0.5q_{x+10}} = \dfrac{4}{96}$. This yields $1000{}_{10.5}V_x = 1000(1.03)^{-1}\left(\frac{4}{96}\right) + (1.03)^{-1}\left(\frac{92}{96}\right)(340.86) = 357.80$.

20. Answer: 0.058

D.6.2   Solutions to Spreadsheet Exercises

1. ${}_{10}V_{30} = 0.09541$ with $i = 4\%$

$\quad {}_{15}V_{30} = 0.11002$ with $i = 6\%$

Guide: Use formula (6.5.4).

2. For $i = 6\%$, $\Pi_3^s = 0.0706$ and $\Pi_0^r = 0.0052$.

$\quad$ For $i = 4\%$, $\Pi_3^s = 0.0791$ and $\Pi_0^r = 0.0052$.

3. (a.) $G_1 = 15.6$

$\quad G_5 = 20.1$

$\quad$ (b.) accumulated gain $= 98.5$

(c.) $i' = 23.24\%$

4. $(S_1)\, \Pi_1, \ldots, \Pi_4 = 231.09$ and $G_2 = 361.03$

(S_2) the present values of gains is $1{,}396.5$

Guide: Set up a table with premiums, benefits, revenues and gains. Use formula (6.9.1) to calculate the gains. Use the solver feature to find Π, so that $_5V = 0$ with $\Pi_1 = \Pi_2 = \Pi_3 = \Pi_4 = \Pi$.

D.7   Multiple Decrements:Solutions

D.7.1   Theory Exercises

1. $_tp_x = \exp\left(-\int_0^t \mu_{x+s}ds\right) = e^{-0.03t}$ $q_{1,x} = \int_0^1 g_1(t)dt = \int_0^1 {}_tp_x\mu_{1,x+t}dt = 0.00985.$

2. Because T is exponential with parameter $\mu = 0.04$,

$$E[T \mid J = 3] = \frac{1}{Pr(J = 3)}\int_0^\infty t\,{}_tp_x\mu_{3,x+t}dt$$

$$= \frac{1}{Pr(J = 3)}\frac{3}{150}\int_0^\infty t\,{}_tp_x dt$$

$$= \frac{1}{Pr(J = 3)}\frac{3}{150}\frac{150}{6}\int_0^\infty t\,{}_tp_x\mu_{x+t}dt$$

$$= \frac{1}{Pr(J = 3)}\frac{1}{2}E[T] = \frac{1}{Pr(J = 3)}\frac{1}{2}\left(\frac{1}{0.04}\right).$$

Also $Pr(J = 3) = \mu_3/\mu = 0.5$, hence $E[T|J = 3] = E[T] = 25$.

3. The present value of future benefits is

$$1000\int_0^\infty \left(3g_1(t) + 2g_2(t) + g_3(t)\right)e^{-t\delta}dt \quad = \quad 1000\int_0^\infty 0.12e^{-0.07t}e^{-0.03t}dt$$
$$= \quad 1200$$

The net annual premium paid continuously, P, satisfies $P\bar{a}_x = 1200$. Since T is exponentially distributed with parameter 0.07, then $\bar{a}_x = 1/(0.07 + 0.03) = 10$. Hence, $P = 120$.

4. $\int_0^t \mu_{x+s}ds = 2t - \log(t+1)$ and so $_tp_x = (t+1)e^{-2t}$. $\int_0^1 {}_tp_x dt = 0.75 - 1.25e^{-2}$.

$$m_x = \frac{1 - 2e^{-2}}{0.75 - 1.25e^{-2}} = 1.25567.$$

5. The NSP is equal to

$$\int_0^2 2\,{}_tp_x\mu_{1,x+t}dt + \int_0^2 {}_tp_x\mu_{2,x+t}dt = \int_0^2 \left(\frac{2t}{20} + \frac{t}{10}\right){}_tp_x dt$$

Since $_tp_x = \exp\left(-\int_0^t \frac{3s}{20}ds\right) = \exp\left(\frac{-3t^2}{40}\right)$, then the NSP is given by

$$\frac{2}{10}\int_0^2 t\exp\left(\frac{-3t^2}{40}\right)dt = \frac{4}{3}\left(1 - e^{-0.3}\right) = 0.3456.$$

6. Apply equation (7.3.6) three times. Use the relation $q_{1,30} + q_{2,30} + q_{3,30} = q_{30}$ to solve for $q_{30} = 0.375$.

7. (a) $10000(0.84)(0.02) = 168$ (b) $10000(1 - 0.01 - 0.25)(0.02) = 148$.

8. $500_{2|}q_{2,63} = 500p_{63}p_{64}q_{2,65}$ Now calculate in order:

(i) $p_{63} = 1 - q_{1,63} - q_{2,63} = 0.45$

(ii) $p_{63}q_{64} = {}_{1|}q_{63} = 0.07$

(iii $q_{64} = \frac{7}{45}$ and ${}_2p_{63} = 0.38$

(iv) ${}_2p_{63}q_{1,65} = 0.042$ so $q_{1,65} = \frac{0.042}{0.38}$

(v) $q_{65} = 1$ and so $q_{2,65} = \frac{0.338}{0.38}$

Hence, $500_{2|}q_{2,63} = 500(0.338) = 169.$

9. $0.02 = \dfrac{q_{1,x}}{1 - 0.5q_x}$ and $q_{1,x} + q_{2,x} = q_x$ so $1000q_x = 19.703.$

10. The NSP is given by

$$\int_0^\infty c_1 v^t {}_tp_x \mu_{1,x+t} dt + \int_0^\infty c_2 v^t {}_tp_x \mu_{2,x+t} dt$$

$$= c_1\mu_1\bar{a}_x + c_2 \int_0^\infty v^t {}_tp_x \left(\mu_{x+t} - \mu_{1,x+t}\right) dt$$

$$= c_1\mu_1\bar{a}_x + c_2\bar{A}_x - c_2\mu_1\bar{a}_x.$$

Hence, the net annual premium is $\text{NSP}/\bar{a}_x = (c_1 - c_2)\,\mu_1 + c_2\bar{P}_x.$

D.8   Multiple Life Insurance: Solutions

D.8.1   Theory Exercises

1. Apply equations (2.6.3) and (2.6.4).

$$
\begin{aligned}
q^1_{80:81} &= \int_0^1 {}_tp_{80:81}\mu_{80+t}dt \\
&= \int_0^1 {}_tp_{80}{}_tp_{81}\left(\mu_{80+t}\right)dt \\
&= \int_0^1 {}_tp_{81}q_{80}dt \\
&= q_{80}(1 - 0.5q_{81}) = 0.3125.
\end{aligned}
$$

Similarly

$$
\begin{aligned}
q^2_{80:81} &= \int_0^1 \left(1 - {}_tp_{80}\right){}_tp_{81}\left(\mu_{81+t}\right)dt \\
&= 0.5q_{80}q_{81} \\
&= 0.1875
\end{aligned}
$$

$$
\begin{aligned}
q_{80:81} &= q^1_{80:81} + q^{\,1}_{80:81} \\
&= q_{80}(1 - 0.5q_{81}) + q_{81}(1 - 0.5q_{80}) \\
&= q_{80} + q_{81} - q_{80}q_{81} \\
&= 0.875
\end{aligned}
$$

$$
\begin{aligned}
q_{\overline{80:81}} &= q_{80}q_{81} \\
&= 0.325.
\end{aligned}
$$

2.

$$
\begin{aligned}
\bar{A}^2_{xy} &= \int_0^\infty v^t(1 - {}_tp_y){}_tp_x\mu_{x+t}dt \\
&= \int_0^\infty e^{-\delta t}(1 - e^{-\mu_x t})e^{-\mu_y t}\mu_x dt \\
&= \mu_x\left(\frac{1}{\delta + \mu_x} - \frac{1}{\delta + \mu_x + \mu_y}\right) \\
&= 0.1167.
\end{aligned}
$$

3. For non-smokers,

$$
{}_tp_x = \frac{75 - x - t}{75 - x} \text{ for } 0 \le t \le 75 - x.
$$

Let $'$ denote the mortality functions for smokers. Since $\mu'_z = 2\mu_z$, $z \geq 0$, then

$$
\begin{aligned}
{}_tp'_x &= \exp\left(-\int_x^{x+t} \mu'_z dz\right) \\
&= \exp\left(-2\int_x^{x+t} \mu_z dz\right) \\
&= ({}_tp_x)^2
\end{aligned}
$$

Hence

$$
\begin{aligned}
\overset{\circ}{e}_{65:55} &= \int_0^{10} {}_tp_{65}\,{}_tp'_{55}\,dt \\
&= \int_0^{10} {}_tp_{65}\left({}_tp_{55}\right)^2 dt \\
&= \int_0^{10} \left(\frac{10-t}{10}\right)\left(\frac{20-t}{20}\right)^2 dt \\
&= 3\frac{13}{24}.
\end{aligned}
$$

4. From the equation obtained from the equivalence principle, we have

$$
10,000\bar{A}_{\overline{xy}} = c\bar{a}_x + 0.5c\left(\bar{a}_y - \bar{a}_{xy}\right).
$$

Use $\bar{A}_{\overline{xy}} = 1 - \delta\bar{a}_{\overline{xy}}$ and $\bar{a}_{\overline{xy}} = \bar{a}_x + \bar{a}_y - \bar{a}_{xy}$ together with the given values to determine $c = 103.45$.

5. Let Π denote the answer. By the equivalence principle,

$$
\Pi\ddot{a}_{xx} = A_{\overline{xx}}.
$$

Since $\ddot{a}_x = 1 + a_x = 10$ and $A_x = 1 - d\ddot{a}_x$, then $d = 0.06$. Since $A_{xx} = 1 - d\ddot{a}_{xx}$, then $\ddot{a}_{xx} = 7.5$. Since $A_{xx} + A_{\overline{xx}} = 2A_x$, $A_{\overline{xx}} = 0.25$. Hence $\Pi = \frac{1}{30}$.

6. The net single premium is

$$
0.5\,\bar{A}^1_{xy} + \bar{A}_{\overline{xy}} = 0.5\,\bar{A}^1_{xy} + \bar{A}_x + \bar{A}_y - \bar{A}_{xy}
$$

Now use the following result from the discussion of Gompertz' Law in section 8.3:

$$
\begin{aligned}
\bar{A}^1_{xy} &= \int_0^\infty v^t\,{}_tp_{xy}\,\mu_{x+t}\,dt \\
&= \int_0^\infty v^t\,{}_tp_{xy}\,Bc^{x+t}\,dt \\
&= \left(\frac{c^x}{c^x + c^y}\right)\int_0^\infty v^t\,{}_tp_{xy}\,Bc^{w+t}\,dt
\end{aligned}
$$

$$= \left(\frac{c^x}{c^x + c^y}\right) \int_0^\infty v^t \, {}_tp_{xy}\mu_{w+t} dt$$

$$= \left(\frac{c^x}{c^x + c^y}\right) \bar{A}_w$$

$$= c^{x-w} \bar{A}_w$$

where $c^w = c^x + c^y$. The solution now follows easily by using $\bar{A}_{xy} = \bar{A}_w$.

7.

$$\infty q^1_{50M:50F} = \int_0^\infty {}_tp_{50M} \, {}_tp_{50F} \mu^F_{50+t} dt$$

$$= \int_0^{50} e^{-0.04t} \frac{1}{50} dt$$

$$= 0.4323.$$

Therefore, $\infty q^2_{50M:50F} = 1 - \infty q^1_{50M:50F} = 0.5677.$

8.

$$E[Z] = \bar{A}^2_{y:x}$$

$$= \int_0^\infty v^t \left(1 - {}_tp_x\right) {}_tp_y \mu_y dt$$

$$= \frac{\mu_y}{\delta + \mu_y} - \frac{\mu_y}{\delta + \mu_y + \mu_x}$$

$$= \frac{21}{110}$$

$$= 0.1909091$$

$$E[Z^2] = {}^2\bar{A}^2_{y:x}$$

$$= \frac{\mu_y}{2\delta + \mu_y} - \frac{\mu_y}{2\delta + \mu_y + \mu_x}$$

$$= \frac{3}{28}$$

$$= 0.1071429$$

$$Var[Z] = 0.0706966$$

9. Let Π denote the initial premium.

$$1000 A_{\overline{25:25}} = 0.4\Pi\ddot{a}_{25:25} + 0.6\Pi\ddot{a}_{\overline{25:25}}$$

$$\Pi = \frac{1000 \left(2A_{25} - A_{25:25}\right)}{1.2\ddot{a}_{25} - 0.2\ddot{a}_{25:25}}$$

$$= 3.5349.$$

10.

$$E[(i)] = \int_0^\infty v^t \left({}_tp_x\right)^{\text{Paul}} \left({}_tp_x\mu_{x+t}\right)^{\text{John}} dt$$

$$
\begin{aligned}
&= \bar{A}^1_{xx} \\
E[(ii)] &= 2E[(i)] \\
E[(iii)] &= 3 \int_0^\infty v^t \left(1 - {}_t p_x\right)^{\text{Paul}} \left({}_t p_x \mu_{x+t}\right)^{\text{John}} dt \\
&= 3 \left(\bar{A}_x - \bar{A}^1_{xx}\right) \\
E[(iv)] &= 4 \left(\bar{A}_x - \bar{A}^1_{xx}\right) \\
\text{Total} &= 7\bar{A}_x - 4 \bar{A}^1_{xx} = 7\bar{A}_x - 2\bar{A}_{xx}.
\end{aligned}
$$

D.8.2 Solutions to Spreadsheet Exercises

1. Check values: $a_{65:60} = 7.9479$, $a_{\overline{65:60}} = 12.7011$, and $a_{65/60} = 3.10736$.

Guide: Set up a table with values for ${}_t p_{60}$ and ${}_t p_{65}$. Use formulas (8.2.3) and (8.7.3) to calculate the corresponding joint and last surviver probabilities. Calulate the annuities based on this probabilities. Use the discrete version of (8.7.6) to calculate the reversionary annuity.

2. The variance is 589.772.

Guide: From the Illustrative Life Table, calculate probabilities of survival for the last-survivor status. See the guide to exercise 1. Find the present value for the annuities certain and calculate the expected present value. Use formula (4.2.9).

3. $P = 0.0078$

Guide: Calculate probabilities of survival for the joint and last-survivor status. Find the insurance single premium for the last survivor status and the annuity for the joint status.

4. $_3V = 0.0258$ $\qquad _7V = 0.2247$

Guide: Use equation (6.3.4) to find the reserve for the first five years. Recognize that after the death of (40), the reserves are the net single premiums for (35).

5. $\bar{a}_{30:40} \approx 7.02017$ and $\bar{A}^1_{30:40} \approx 0.14558$

Guide: For a Makeham law, the formula (7.3.4) provides ${}_t p_x$. Integrate numerically to find $\bar{a}_x$ for typical values of x, A, B and c in four cells. Now use the fact that $\mu_{30:40}(t) = A' + Bc^{w+t} = \mu_{w+t}$ where $A' = 2A$ and $c^w = c^x + c^y$, so $\bar{a}_{30:40} = \bar{a}_w$ with these values in the appropriate cells: $x = \log\left(c^{30} + c^{40}\right) / \log c = 41.58$, $c = 1.15$, $A = 0.008$, and $B = 0.0001$. To obtain $\bar{A}^1_{30:40}$ either evaluate an integral numerically or use the relations:

$$
\bar{A}^1_{xy} = \frac{c^x}{c^w}\left(\bar{A}_{xy} - A(1 - c^{y-x})\bar{a}_{xy}\right) \quad \text{and} \quad \bar{A}_{xy} = 1 - \delta\bar{a}_{xy}
$$

D.9   The Total Claim Amount in a Portfolio

D.9.1   Theory Exercises

1. $E[S_1] = 1000(0.001) + 100(0.15) = 16$ and $Var[S_1] = 1000^2(0.001) + 100^2(0.15) - 16^2 = 2244$. $E[S] = 5E[S^1] = 80$ and $Var[S] = 5Var[S_1] = 11,220$. $Pr(S > 200) = 1 - \Phi((200 - 80)/\sqrt{(11,220))}) \approx 1 - \Phi(1.1329) \approx 0.1286$. The exact value is $Pr(S > 200) = 1 - Pr(S = 0) - Pr(S = 100) - Pr(S = 200) = 1 - (0.849)^5 - 5(0.15)(0.849)^4 - 10(0.15)^2(0.849)^3 \approx 0.032$.

2. $\rho(\beta) = E[(S - \beta)^+] = \sigma \int_k^\infty (x - k)\phi(x)dx$ where $\phi(x) = e^{-x^2/2}/\sqrt{2\pi}$ and $k = (\beta - \mu)/\sigma$. Hence $\rho(\beta) = \sigma \int_k^\infty x\phi(x)dx - \sigma k(1 - \Phi(k))$. Now use the fact that $x\phi(x) = -\phi'(x)$ to obtain

$$\rho(\beta) = \sigma\phi(k) - \sigma k(1 - \Phi(k)) = \sigma\phi(-k) + (\mu - \beta)\Phi(-k)$$

$$= \sigma\phi(\frac{\beta - \mu}{\sigma}) + (\mu - \beta)\Phi(\frac{\mu - \beta}{\sigma}).$$

3. $M_S(t) = \sum_{k=0}^\infty E[\exp(t(X_1 + \cdots + X_k))] Pr(N = k) = \sum_{k=0}^\infty (M_X(t))^k Pr(N = k)$

$= \sum_{k=0}^\infty \exp(k\log(M_X(t))) Pr(N = k) = M_N(\log(M_X(t)))$

Differentiate to get the moment relations:

$M_S'(t) = M_N'(\log(M_X(t)))\dfrac{M_X'(t)}{M_X(t)}$ and

$M_S''(t) = M_N''(\log(M_X(t)))\left(\dfrac{M_X'(t)}{M_X(t)}\right)^2$

$\quad + \ M_N'(\log(M_x(t)))\left[\dfrac{M_X''(t)M_X(t) - (M_X'(t))^2}{M_x(t)^2}\right]$ Evaluate with $t = 0$.

4. $E[R] = \int_\beta^\gamma [1 - F(x)]dx$

5. (a) $F(x) = 1 + \rho'(x), x \geq 0$

(b) $F(x) = 1 - \left(1 + \dfrac{1}{2}x + \dfrac{1}{4}x^2\right)e^{-x}$ and $f(x) = \left(\dfrac{1}{2} + \dfrac{1}{4}x^2\right)e^{-x}$

6. e^{-6}

7. $\rho(\beta) = e^{\mu + \sigma^2/2}\left[1 - \Phi\left(\dfrac{\log\beta - \mu - \sigma^2}{\sigma}\right)\right] - \beta\left[1 - \Phi\left(\dfrac{\log\beta - \mu}{\sigma}\right)\right]$

8. (a)

x	$Pr(S_1 + S_2 = x)$	$Pr(S_1 + S_2 + S_3 = x)$	$F(x)$
0	0.56	0.336	0.336
0.5	0.07	0.042	0.378
1.0	0.08	0.048	0.426
1.5	0.09	0.166	0.592
2.0	0.01	0.020	0.612
2.5	0.15	0.162	0.774
3.0	0.01	0.087	0.861
3.5	0.01	0.023	0.884
4.0	0.01	0.053	0.937
4.5		0.012	0.949
5.0	0.01	0.024	0.973
5.5		0.018	0.991
6.0		0.002	0.993
6.5		0.004	0.997
7.0		0.001	0.998
7.5		0.001	0.999
8.0		0.001	1.000

(b)

x	$f(x)$	$F(x)$
0	0.4066	0.4066
0.5	0.0407	0.4472
1.0	0.0427	0.4899
1.5	0.1261	0.6160
2.0	0.1364	0.7524
2.5	0.0659	0.8183
3.0	0.0365	0.8548
3.5	0.0446	0.8994
4.0	0.0372	0.9366
4.5	0.0214	0.9580
5.0	0.0126	0.9707
5.5	0.0101	0.9807
6.0	0.0075	0.9882
6.5	0.0045	0.9927
7.0	0.0026	0.9954
7.5	0.0018	0.9971
8.0	0.0012	0.9983

D.10 Expense Loadings

D.10.1 Theory Exercises

1.a.

Development of Endowment Reserves

k	Net Premium	Expense-Loaded
0	0	
1	77	31
2	158	116
3	244	206
4	335	302
5	431	403
6	532	509
7	639	621
8	752	740
9	873	867

1.b. 83.60

2. $-_kV^\alpha$ is the unamortized acquisition expense at the end of policy year k.

3.a. The expense-loaded premium is 22.32

Development of Term Insurance Reserves

k	Net Premium	Expense-Loaded
0	0.0	
1	1.3	-35.6
2	2.3	-31.3
3	3.1	-27.1
4	3.7	-22.9
5	4.0	-18.8
6	3.9	-14.8
7	3.6	-10.9
8	2.8	-7.10
9	1.6	-3.50

3.b. The one-year net cost of insurance is $1000vq_x = 16.03$. The first year loading is $22.32 - 16.03 = 6.29$. The acquisition expense is 40, requiring an investment of $40 - 6.29 = 33.71$.

4. $1000P = 11.06, 1000P^\alpha = 0.78, 1000P^\beta = 2.28$, and $1000P^\gamma = 1.13$. The expense-loaded premium is $1000P^a = 15.25$.

5. The first 10 years of reserves are given below. They can be developed easily using a spreadsheet program and the recursion relations:

k	q_{x+k}	1000_kV	1000_kV^α	1000_kV^γ	1000_kV^a
0	0.00201	0.00	0.00	0.00	0.00
1	0.00214	9.63	-11.81	0.13	-2.05
2	0.00228	19.62	-11.61	0.27	8.29
3	0.00243	30.01	-11.40	0.42	19.03
4	0.00260	40.80	-11.18	0.58	30.19
5	0.00278	51.99	-10.95	0.74	41.78
6	0.00298	63.60	-10.71	0.91	53.80
7	0.00320	75.64	-10.47	1.10	66.27
8	0.00344	88.12	-10.21	1.29	79.20
9	0.00371	101.05	-9.94	1.49	92.61
10	0.00400	114.43	-9.65	1.71	106.49

$$_kV^\gamma = \begin{cases} vp_{x+k}\ _{k+1}V^\gamma + \gamma - P^\gamma & \text{for } 0 \le k \le 29 \\ vp_{x+k}\ _{k+1}V^\gamma + \gamma & \text{for } 30 \le k \le 64 \\ \gamma & \text{for } k = 64 \end{cases}$$

$$_kV^\alpha = \begin{cases} vp_{x+k}\ _{k+1}V^\alpha - P^\alpha & \text{for } 0 \le k \le 29 \\ 0 & \text{for } k \ge 30 \end{cases}$$

$$_kV^a = \begin{cases} v\left(p_{x+k}\ _{k+1}V^a + 1000q_{x+k}\right) - (1-\beta)P^a + \gamma & \text{for } 0 \le k \le 29 \\ v\left(p_{x+k}\ _{k+1}V^a + 1000q_{x+k}\right) + \gamma & \text{for } 30 \le k \le 64 \\ \gamma + 1000v & \text{for } k = 64 \end{cases}$$

D.10.2   Spreadsheet Exercises

1. Guide: Use the recursion formulas. Here are the results: $1000P = 28.42, 1000P^\alpha = 1.70, 1000P^\beta = 1.74$, and $1000P^\gamma = 3.0$. The expense-loaded premium is $1000P^a = 34.68$.

k	q_{x+k}	1000_kV	1000_kV^α	1000_kV^γ	1000_kV^a
0	0.00278	0.00	-20.00	0.00	-20.00
1	0.00298	27.42	-19.45	0.00	7.97
2	0.00320	56.38	-18.87	0.00	37.51
3	0.00344	86.97	-18.26	0.00	68.71
4	0.00371	119.28	-17.61	0.00	101.67
5	0.00400	153.42	-16.93	0.00	136.49
6	0.00431	189.51	-16.21	0.00	173.30
7	0.00466	227.68	-15.45	0.00	212.23
8	0.00504	268.06	-14.64	0.00	253.42
9	0.00546	310.79	-13.78	0.00	297.01
10	0.00592	356.05	-12.88	0.00	343.17
11	0.00642	404.01	-11.92	0.00	392.09
12	0.00697	454.88	-10.90	0.00	443.98
13	0.00758	508.87	-9.82	0.00	499.05
14	0.00824	566.24	-8.68	0.00	557.57
15	0.00896	627.27	-7.45	0.00	619.82
16	0.00975	692.28	-6.15	0.00	686.12
17	0.01062	761.62	-4.77	0.00	756.85
18	0.01158	835.69	-3.29	0.00	832.41
19	0.01262	914.98	-1.70	0.00	913.28

D.11 Estimating Probabilities of Death

D.11.1 Theory Exercises

1. (a) (0.00553, 0.00981)
   (b) (0.00682, 0.00811)

2. 0.007388, 0.00112, 0.007360

3.a.

$$\sum_{k=n}^{\infty} \frac{(\lambda^l)^k}{k!} e^{-\lambda^l} = w$$

or

$$\sum_{k=0}^{n-1} \frac{(\lambda^l)^k}{k!} e^{-\lambda^l} = 1 - w$$

b.

$$\sum_{k=0}^{n} \frac{(\lambda^u)^k}{k!} e^{-\lambda^u} = w$$

c.

$$\int_{\lambda^l}^{\infty} f(x;n)dx = 1 - w$$

$$\int_{\lambda^u}^{\infty} f(x;n+1)dx = w$$

4. 0.7326, 1.365 years

5. 0.0346

6. The classical estimator is $4/145 \approx 0.02759$. The MLE is $1 - \exp(-4/143) \approx 0.02758$.

7. $\hat{\mu} = 1/9.1$

8. $\hat{\mu} = \dfrac{90}{1000}, \hat{q} = 1 - \bar{e}^{\hat{\mu}} = 0.0861$.

$$\hat{q}_1 = \frac{4}{9} \hat{q} = 0.0383$$

$$\hat{q}_2 = \frac{5}{9} \hat{q} = 0.0478$$

9. Observed deaths $= 45$. Expected deaths $= 19.4$ from the Illustrative Life Table. From the table in section 11.5, we get a 90% confidence interval $34.56 \le \lambda \le 57.69$. Therefore $\hat{f} = 45/19.4 = 2.32$ and the 90% interval is $(1.78, 2.97)$.

Appendix E

Tables

E.0 Illustrative Life Tables

Basic Functions and Net Single Premiums at $i = 5\%$ [1]

x	ℓ_x	d_x	$1000q_x$	$\ddot{a}_x$	$1000A_x$	$1000\left(^2A_x\right)$	x
0	10,000,000	204,200	20.42	19.642724	64.63	28.72	0
1	9,795,800	13,126	1.34	19.982912	48.43	11.47	1
2	9,782,674	11,935	1.22	19.958801	49.58	11.33	2
3	9,770,739	10,943	1.12	19.931058	50.90	11.28	3
4	9,759,796	10,150	1.04	19.899898	52.39	11.33	4
5	9,749,646	9,555	0.98	19.865552	54.02	11.46	5
6	9,740,091	9,058	0.93	19.828262	55.80	11.67	6
7	9,731,033	8,661	0.89	19.788078	57.71	11.95	7
8	9,722,372	8,458	0.87	19.745056	59.76	12.29	8
9	9,713,914	8,257	0.85	19.699446	61.93	12.69	9
10	9,705,657	8,250	0.85	19.651122	64.23	13.15	10
11	9,697,407	8,243	0.85	19.600339	66.65	13.66	11
12	9,689,164	8,333	0.86	19.546971	69.19	14.23	12
13	9,680,831	8,422	0.87	19.491083	71.85	14.84	13
14	9,672,409	8,608	0.89	19.432543	74.64	15.50	14
15	9,663,801	8,794	0.91	19.371410	77.55	16.21	15
16	9,655,007	8,979	0.93	19.307550	80.59	16.98	16
17	9,646,028	9,164	0.95	19.240821	83.77	17.81	17
18	9,636,864	9,348	0.97	19.171075	87.09	18.70	18
19	9,627,516	9,628	1.00	19.098154	90.56	19.67	19
20	9,617,888	9,906	1.03	19.022085	94.19	20.71	20
21	9,607,982	10,184	1.06	18.942699	97.97	21.82	21
22	9,597,798	10,558	1.10	18.859825	101.91	23.02	22
23	9,587,240	10,929	1.14	18.773468	106.03	24.31	23
24	9,576,311	11,300	1.18	18.683440	110.31	25.69	24
25	9,565,011	11,669	1.22	18.589547	114.78	27.17	25
26	9,553,342	12,133	1.27	18.491584	119.45	28.77	26
27	9,541,209	12,690	1.33	18.389518	124.31	30.49	27
28	9,528,519	13,245	1.39	18.283311	129.37	32.33	28
29	9,515,274	13,892	1.46	18.172737	134.63	34.30	29
30	9,501,382	14,537	1.53	18.057738	140.11	36.41	30
31	9,486,845	15,274	1.61	17.938070	145.81	38.67	31
32	9,471,571	16,102	1.70	17.813654	151.73	41.09	32
33	9,455,469	16,925	1.79	17.684401	157.89	43.68	33
34	9,438,544	17,933	1.90	17.550034	164.28	46.45	34
35	9,420,611	18,935	2.01	17.410616	170.92	49.40	35
36	9,401,676	20,120	2.14	17.265850	177.82	52.56	36
37	9,381,556	21,390	2.28	17.115771	184.96	55.93	37
38	9,360,166	22,745	2.43	16.960229	192.37	59.52	38
39	9,337,421	24,277	2.60	16.799062	200.04	63.35	39
40	9,313,144	25,891	2.78	16.632259	207.99	67.41	40
41	9,287,253	27,676	2.98	16.459630	216.21	71.74	41
42	9,259,577	29,631	3.20	16.281129	224.71	76.34	42
43	9,229,946	31,751	3.44	16.096696	233.49	81.23	43
44	9,198,195	34,125	3.71	15.906249	242.56	86.41	44
45	9,164,070	36,656	4.00	15.709844	251.91	91.90	45
46	9,127,414	39,339	4.31	15.507365	261.55	97.71	46
47	9,088,075	42,350	4.66	15.298671	271.49	103.86	47
48	9,045,725	45,590	5.04	15.083893	281.72	110.36	48
49	9,000,135	49,141	5.46	14.862997	292.24	117.23	49

Basic Functions and Net Single Premiums at $i = 5\%$

x	ℓ_x	d_x	$1000q_x$	$\ddot{a}_x$	$1000A_x$	$1000\left({}^2A_x\right)$	x
50	8,950,994	52,990	5.92	14.636061	303.04	124.46	50
51	8,898,004	57,125	6.42	14.403131	314.14	132.08	51
52	8,840,879	61,621	6.97	14.164220	325.51	140.10	52
53	8,779,258	66,547	7.58	13.919449	337.17	148.53	53
54	8,712,711	71,793	8.24	13.669034	349.09	157.36	54
55	8,640,918	77,423	8.96	13.413011	361.29	166.63	55
56	8,563,495	83,494	9.75	13.151498	373.74	176.33	56
57	8,480,001	90,058	10.62	12.884699	386.44	186.47	57
58	8,389,943	97,156	11.58	12.612883	399.39	197.05	58
59	8,292,787	104,655	12.62	12.336384	412.55	208.08	59
60	8,188,132	112,669	13.76	12.055342	425.94	219.56	60
61	8,075,463	121,213	15.01	11.770064	439.52	231.49	61
62	7,954,250	130,291	16.38	11.480898	453.29	243.87	62
63	7,823,959	139,892	17.88	11.188205	467.23	256.69	63
64	7,684,067	149,993	19.52	10.892369	481.32	269.94	64
65	7,534,074	160,626	21.32	10.593780	495.53	283.63	65
66	7,373,448	171,728	23.29	10.292913	509.86	297.73	66
67	7,201,720	183,212	25.44	9.990230	524.27	312.23	67
68	7,018,508	195,044	27.79	9.686158	538.75	327.11	68
69	6,823,464	207,229	30.37	9.381167	553.28	342.37	69
70	6,616,235	219,527	33.18	9.075861	567.82	357.96	70
71	6,396,708	231,945	36.26	8.770664	582.35	373.88	71
72	6,164,763	244,248	39.62	8.466182	596.85	390.09	72
73	5,920,515	256,358	43.30	8.162908	611.29	406.56	73
74	5,664,157	267,971	47.31	7.861452	625.65	423.26	74
75	5,396,186	278,929	51.69	7.562297	639.89	440.15	75
76	5,117,257	288,972	56.47	7.265989	654.00	457.21	76
77	4,828,285	297,809	61.68	6.973063	667.95	474.38	77
78	4,530,476	305,218	67.37	6.683979	681.71	491.66	78
79	4,225,258	310,810	73.56	6.399299	695.27	508.97	79
80	3,914,448	314,330	80.30	6.119411	708.60	526.31	80
81	3,600,118	315,514	87.64	5.844708	721.68	543.60	81
82	3,284,604	314,041	95.61	5.575590	734.50	560.84	82
83	2,970,563	309,770	104.28	5.312277	747.03	577.99	83
84	2,660,793	302,506	113.69	5.055031	759.28	594.95	84
85	2,358,287	292,168	123.89	4.803941	771.24	611.84	85
86	2,066,119	278,802	134.94	4.558938	782.91	628.49	86
87	1,787,317	262,539	146.89	4.319797	794.29	645.05	87
88	1,524,778	243,675	159.81	4.085991	805.43	661.36	88
89	1,281,103	222,592	173.75	3.856599	816.35	677.76	89
90	1,058,511	199,815	188.77	3.630178	827.13	694.07	90
91	858,696	175,973	204.93	3.404320	837.89	710.54	91
92	682,723	151,749	222.27	3.175241	848.80	727.33	92
93	530,974	127,890	240.86	2.936743	860.15	745.41	93
94	403,084	105,096	260.73	2.678823	872.44	764.65	94
95	297,988	84,006	281.91	2.384461	886.46	787.86	95
96	213,982	74,894	350.00	2.024369	903.60	817.60	96
97	139,088	66,067	475.00	1.654770	921.21	847.96	97
98	73,021	49,289	675.00	1.309539	937.65	885.10	98
99	23,732	23,732	1000.00	1.000000	952.35	914.67	99

E.1 Commutation Columns

Illustrative Life Table and $i = 5\%$

x	D_x	N_x	C_x	M_x	x
0	10,000,000.0	196,427,244.2	194,476.190	646,321.725	0
1	9,329,333.3	186,427,244.2	11,905.669	451,845.535	1
2	8,873,173.7	177,097,910.9	10,309.902	439,939.866	2
3	8,440,331.7	168,224,737.2	9,002.833	429,629.964	3
4	8,029,408.3	159,784,405.5	7,952.791	420,627.131	4
5	7,639,102.8	151,754,997.2	7,130.088	412,674.340	5
6	7,268,205.9	144,115,894.4	6,437.351	405,544.252	6
7	6,915,663.5	136,847,688.5	5,862.106	399,106.901	7
8	6,580,484.1	129,932,025.0	5,452.102	393,244.795	8
9	6,261,675.6	123,351,540.9	5,069.082	387,792.693	9
10	5,958,431.5	117,089,865.3	4,823.604	382,723.611	10
11	5,669,873.0	111,131,433.8	4,590.011	377,900.007	11
12	5,395,289.1	105,461,560.8	4,419.168	373,309.996	12
13	5,133,951.4	100,066,271.7	4,253.682	368,890.828	13
14	4,885,223.8	94,932,320.3	4,140.595	364,637.146	14
15	4,648,453.5	90,047,096.5	4,028.633	360,496.551	15
16	4,423,070.0	85,398,643.0	3,917.508	356,467.918	16
17	4,208,530.1	80,975,573.0	3,807.831	352,550.410	17
18	4,004,316.0	76,767,042.9	3,699.321	348,742.579	18
19	3,809,935.0	72,762,726.9	3,628.692	345,043.258	19
20	3,624,880.8	68,952,791.9	3,555.683	341,414.566	20
21	3,448,711.8	65,327,911.1	3,481.399	337,858.883	21
22	3,281,006.0	61,879,199.3	3,437.382	334,377.484	22
23	3,121,330.2	58,598,193.3	3,388.732	330,940.102	23
24	2,969,306.7	55,476,863.1	3,336.921	327,551.370	24
25	2,824,574.3	52,507,556.4	3,281.798	324,214.449	25
26	2,686,788.9	49,682,982.1	3,249.804	320,932.651	26
27	2,555,596.8	46,996,193.2	3,237.138	317,682.847	27
28	2,430,664.6	44,440,596.4	3,217.824	314,445.709	28
29	2,311,700.8	42,009,931.8	3,214.296	311,227.885	29
30	2,198,405.5	39,698,231.0	3,203.366	308,013.589	30
31	2,090,516.2	37,499,825.5	3,205.496	304,810.223	31
32	1,987,762.3	35,409,309.3	3,218.348	301,604.727	32
33	1,889,888.6	33,421,547.0	3,221.755	298,386.379	33
34	1,796,672.2	31,531,658.4	3,251.079	295,164.624	34
35	1,707,865.3	29,734,986.2	3,269.268	291,913.545	35
36	1,623,269.1	28,027,120.9	3,308.445	288,644.277	36
37	1,542,662.1	26,403,851.8	3,349.789	285,335.832	37
38	1,465,852.2	24,861,189.7	3,392.370	281,986.043	38
39	1,392,657.4	23,395,337.5	3,448.443	278,593.673	39
40	1,322,891.9	22,002,680.1	3,502.576	275,145.230	40
41	1,256,394.5	20,679,788.2	3,565.765	271,642.654	41
42	1,193,000.4	19,423,393.7	3,635.854	268,076.889	42
43	1,132,555.0	18,230,393.3	3,710.464	264,441.035	43
44	1,074,913.3	17,097,838.3	3,797.993	260,730.571	44
45	1,019,929.0	16,022,925.0	3,885.414	256,932.578	45
46	967,475.5	15,002,996.0	3,971.241	253,047.164	46
47	917,434.0	14,035,520.5	4,071.618	249,075.923	47
48	869,675.1	13,118,086.5	4,174.399	245,004.305	48
49	824,087.6	12,248,411.4	4,285.278	240,829.906	49

Illustrative Life Table and $i = 5\%$

x	D_x	N_x	C_x	M_x	x
50	780,560.0	11,424,323.8	4,400.881	236,544.628	50
51	738,989.6	10,643,763.8	4,518.379	232,143.747	51
52	699,281.3	9,904,774.2	4,641.901	227,625.368	52
53	661,340.3	9,205,492.9	4,774.263	222,983.467	53
54	625,073.6	8,544,152.6	4,905.357	218,209.204	54
55	590,402.8	7,919,079.0	5,038.129	213,303.847	55
56	557,250.3	7,328,676.2	5,174.462	208,265.718	56
57	525,540.1	6,771,425.9	5,315.486	203,091.256	57
58	495,198.9	6,245,885.8	5,461.362	197,775.770	58
59	466,156.6	5,750,686.9	5,602.760	192,314.408	59
60	438,355.9	5,284,530.3	5,744.566	186,711.648	60
61	411,737.3	4,846,174.4	5,885.897	180,967.082	61
62	386,244.8	4,434,437.1	6,025.437	175,081.185	62
63	361,826.8	4,048,192.3	6,161.376	169,055.748	63
64	338,435.6	3,686,365.5	6,291.679	162,894.372	64
65	316,027.9	3,347,929.9	6,416.853	156,602.693	65
66	294,562.1	3,031,902.0	6,533.683	150,185.840	66
67	274,001.7	2,737,339.9	6,638.678	143,652.157	67
68	254,315.3	2,463,338.2	6,730.866	137,013.479	68
69	235,474.2	2,209,022.9	6,810.823	130,282.613	69
70	217,450.3	1,973,548.7	6,871.439	123,471.790	70
71	200,224.1	1,756,098.4	6,914.416	116,600.351	71
72	183,775.2	1,555,874.3	6,934.453	109,685.935	72
73	168,089.5	1,372,099.1	6,931.684	102,751.482	73
74	153,153.6	1,204,009.6	6,900.656	95,819.798	74
75	138,959.9	1,050,856.0	6,840.801	88,919.142	75
76	125,502.0	911,896.1	6,749.626	82,078.341	76
77	112,776.0	786,394.1	6,624.796	75,328.715	77
78	100,781.0	673,618.1	6,466.295	68,703.919	78
79	89,515.6	572,837.1	6,271.206	62,237.624	79
80	78,981.7	483,321.5	6,040.218	55,966.418	80
81	69,180.5	404,339.8	5,774.257	49,926.200	81
82	60,111.9	335,159.3	5,473.619	44,151.943	82
83	51,775.8	275,047.4	5,142.073	38,678.324	83
84	44,168.2	223,271.6	4,782.374	33,536.251	84
85	37,282.6	179,103.4	4,398.990	28,753.877	85
86	31,108.3	141,820.8	3,997.853	24,354.887	86
87	25,629.1	110,712.5	3,585.383	20,357.034	87
88	20,823.2	85,083.4	3,169.300	16,771.651	88
89	16,662.4	64,260.2	2,757.228	13,602.351	89
90	13,111.7	47,597.8	2,357.230	10,845.123	90
91	10,130.1	34,486.1	1,977.109	8,487.893	91
92	7,670.6	24,356.0	1,623.757	6,510.784	92
93	5,681.6	16,685.4	1,303.294	4,887.027	93
94	4,107.7	11,003.8	1,020.006	3,583.733	94
95	2,892.1	6,896.1	776.493	2,563.727	95
96	1,977.9	4,004.0	659.303	1,787.234	96
97	1,224.4	2,026.1	553.902	1,127.931	97
98	612.2	801.7	393.559	574.029	98
99	189.5	189.5	180.470	180.470	99

E.2 Multiple Decrement Tables

Illustrative Service Table

x	ℓ_x	$d_{1,x}$	$d_{2,x}$	$d_{3,x}$	$d_{4,x}$
30	100,000	100	19,990	0	0
31	79,910	80	14,376	0	0
32	65,454	72	9,858	0	0
33	55,524	61	5,702	0	0
34	49,761	60	3,971	0	0
35	45,730	64	2,693	46	0
36	42,927	64	1,927	43	0
37	40,893	65	1,431	45	0
38	39,352	71	1,181	47	0
39	38,053	72	989	49	0
40	36,943	78	813	52	0
41	36,000	83	720	54	0
42	35,143	91	633	56	0
43	34,363	96	550	58	0
44	33,659	104	505	61	0
45	32,989	112	462	66	0
46	32,349	123	421	71	0
47	31,734	133	413	79	0
48	31,109	143	373	87	0
49	30,506	156	336	95	0
50	29,919	168	299	102	0
51	29,350	182	293	112	0
52	28,763	198	259	121	0
53	28,185	209	251	132	0
54	27,593	226	218	143	0
55	27,006	240	213	157	0
56	26,396	259	182	169	0
57	25,786	276	178	183	0
58	25,149	297	148	199	0
59	24,505	316	120	213	0
60	23,856	313	0	0	3,552
61	19,991	298	0	0	1,587
62	18,106	284	0	0	2,692
63	15,130	271	0	0	1,350
64	13,509	257	0	0	2,006
65	11,246	204	0	0	4,448
66	6,594	147	0	0	1,302
67	5,145	119	0	0	1,522
68	3,504	83	0	0	1,381
69	2,040	49	0	0	1,004
70	987	17	0	0	970

References

A thorough introduction into the theory of compound interest is given in Butcher-Nesbitt [3]. The textbook by Bowers-Gerber-Hickman-Jones-Nesbitt [2] is the natural reference for Chapters 2–10; it contains numerous examples and exercises. The "classical method" of Chapter 11 is documented in Batten [1] and updated in Hoem [7]. In this respect the reader may also orient himself with the text of Elandt-Johnson [6].

An extensive bibliography is given by Wolthuis-van Hoek [14].

The classical texts in life insurance mathematics are those by Zwinggi [15], Saxer [12] and Jordan [9]; the newer book by Wolff [13] is of impressive completeness. The monographs by Isenbart-Münzner [8] and Neill [10] are written in the traditional style; however, the former book may appeal to non-mathematicians. The three volumes by Reichel [11] have an unconventional approach and will appeal to the mathematically minded reader. The books by De Vylder [4] and De Vylder-Jaumain [5] give a very elegant presentation of the subject.

1. Batten, R.W.: Mortality Table Construction. Prentice-Hall, Englewood Cliffs, New Jersey, 1978

2. Bowers, N.L., Gerber, H.U., Hickman, J.C., Jones, D.A., Nesbitt, C.J.: Actuarial Mathematics. Society of Actuaries, Itasca, Illinois, 1986

3. Butcher, M.V., Nesbitt, C.J.: Mathematics of Compound Interest. Ulrich's Books, Ann Arbor, Michigan, 1971

4. De Vylder, Fl.: Théorie générale des opérations d'assurances individuelles de capitalisation. Office des Assureurs de Belgique, Bruxelles, 1973

5. De Vylder, Fl.,Jaumain, C.: Exposé moderne de la théorie mathématique des opérations viagéres. Office des Assureurs de Belgique, Bruxelles, 1976

6. Elandt-Johnson, R.C., Johnson, N.L.: Survival Models and Data Analysis. John Wiley & Sons, New York London Sydney, 1980

7. Hoem, J.M.: A Flaw in Actuarial Exposed-to-Risk Theory. Scandinavian Actuarial Journal 1984(3), 187–194

8. Isenbart, F., Münzner, H.: Lebensversicherungsmathematik für Praxis und Studium. Gabler, Wiesbaden, 1977

9. Jordan, C.W.: Life Contingencies. Second edition. Society of Actuaries, Chicago, Ill., 1967. Also available in braille

10. Neill, A.: Life Contingencies. William Heinemann, London, 1977

11. Reichel, G.: Mathematische Grundlagen der Lebensversicherung. Volumes 3, 5 and 9 of the series Angewandte Versicherungsmathematik der DGVM. Verlag Versicherungswirtschaft E.V., Karlruhe, 1975, 1976, 1978

12. Saxer, W.: Versicherungsmathematik. Volumes 1 and 2. Springer, Berlin Göttingen Heidelberg, 1955, 1958

13. Wolff, K.-H.: Versicherungsmathematik. Springer, Wien New York, 1970

14. Wolthuis, H., van Hoek, I.: Stochastic Models for Life Contingencies. Insurance: Mathematics & Economics, 5, 1986(3), 217–254.

15. Zwinggi, E.: Versicherungsmathematik. Second edition. Birkhäuser, Basel Stuttgart, 1958

Index

G. Ottaviani (Ed.)

Financial Risk in Insurance

1995. XI, 112 pages, 20 figures.
Hardcover DM 120,–
ISBN 3- 540-57054-3

G. Ottaviani (Ed.)

Financial Risk in Insurance

Springer

This book, published with the contribution of INA (National Insurance Institute), contains the "invited contributions" presented at the 3rd International AFIR Colloquium, held in Rome 1993. The colloquium was aimed at encouraging research on the theoretical bases of actuarial sciences, with reference to the standpoints of the theory of finance and of corporate finance, together with mathematical methods, probability and the theory of stochastic processes. In the spirit of actuarial tradition, attention was given to the link between the theoretical approach and the operative problems of financial markets and institutions, and insurance ones in particular.
The book is an important reference work for students and researchers of actuarial sciences but also for professionals.

Please order by
Fax: +49 - 30 - 827 87- 301
e-mail: orders@springer.de
or through your bookseller

Springer

Springer-Verlag, P. O. Box 31 13 40, D-10643 Berlin, Germany.

Printing: COLOR-DRUCK DORFI GmbH, Berlin
Binding: Buchbinderei Lüderitz & Bauer, Berlin